Introductory Calculus

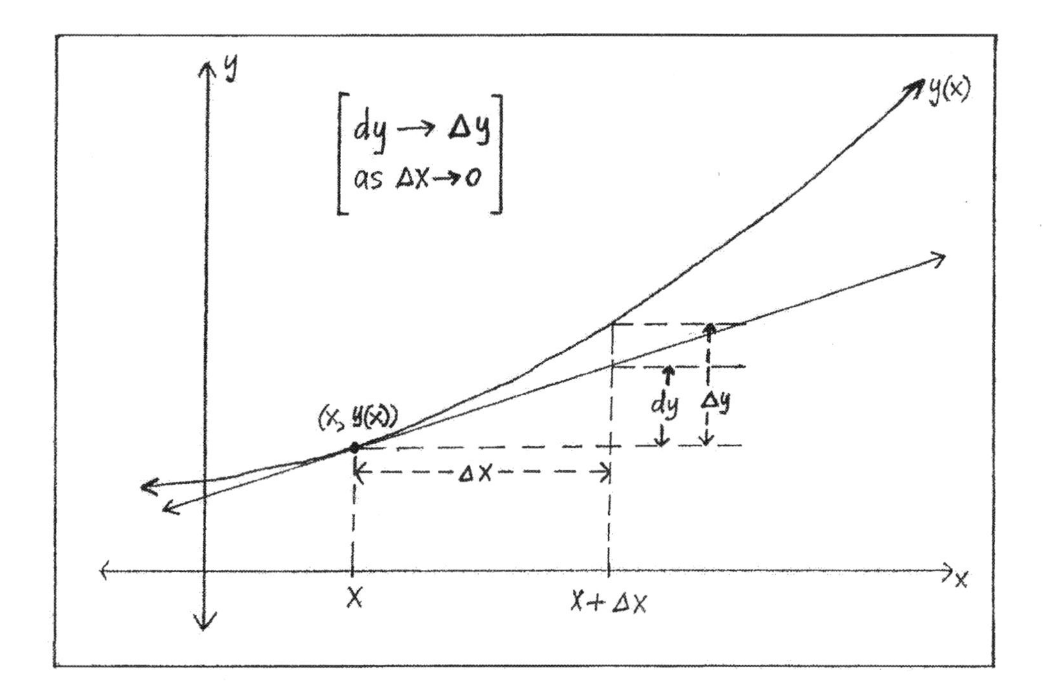

Timothy C. Kearns

authorHOUSE®

AuthorHouse™
1663 Liberty Drive
Bloomington, IN 47403
www.authorhouse.com
Phone: 1 (800) 839-8640

Published by AuthorHouse 11/10/2015

ISBN: 978-1-5049-5925-4 (sc)
ISBN: 978-1-5049-5933-9 (e)

Print information available on the last page.

To my Brother John,
Who urges me to Believe in Myself !

Table of Contents

Part II The Differential Calculus

Part III The Integral Calculus

(8) Indefinite Integrals
- (#) Antiderivatives
- (#) Indefinite Integrals
- (#) Integral Formulas
- (#) Integration Using a Change of Variable
- (#) Integration By Parts
- (#) Partial Fraction Decomposition

(9) The Definite Integral
- (#) Definite Integrals
- (#) Riemann Sums
- (#) The Fundamental Theorem of Calculus

(10) Applications of Indefinite and Definite Integrals
- (#) Areas Between Curves
- (#) Some Applications to Physics
- (#) Arc Length
- (#) Surfaces of Revolution
- (#) Solids of Revolution

(11) Improper Integrals
- (#) Improper Integrals of Two Types

Part IV Differential Equations

Author's Preface

I started studying Calculus in 1979 at the age of 18 when a freshman at Virginia Tech. I immediately loved the subject, and mathematics has been very important to me ever since. I was actually a little disappointed when in my sophomore year Calculus went from being a 5 quarter hour credit course to a 3 quarter hour credit course. I really wanted to be studying Calculus everyday of the week! I progressed through the usual courses in linear algebra, differential equations, advanced calculus, real analysis, plus a smattering of other courses. For the past 12 years I have been tutoring predominately college level students in math and statistics, which is a truly great way to learn mathematics! This book is for the most part right off the top of my head. I had to refresh my memory with the form of the logistic differential equation from the internet source wikipedia.org, but the solution that I have presented is my own (I'm sure that it is very much like any other solution out there). The material in this book and a lot of my knowledge of mathematics has over the years become a part of my soul.

Most people go through this world knowing very little about mathematics, yet it is so important to our rapidly growing technological civilization, and to any serious understanding of the Cosmos we find ourselves in. You too, as a student aspiring to a technical degree, will need to learn math to some level, usually to the level of Calculus, and in many cases beyond. Even 300-350 years after its development, Calculus remains a seminal course in the mathematical training of many college students.

It is important for a student that needs a firm foundation in mathematics to do it right! For your college mathematics courses, your professors will undoubtedly require you to purchase expensive and massive textbooks, which if your intentions are serious, you should definitely get into the habit of reading. This is very important, along with solving as many of your assigned exercises as you can. Mathematics courses are generally difficult, and Calculus is no exception. They require time and an honest effort. The only way to learn mathematics is by doing mathematics. If you follow my advice, your efforts will pay off handsomely as you progress to a degree and beyond.

This book is an inexpensive and relatively compact introduction to Calculus, with plenty of examples illustrating the concepts and calculational techniques that you need to know along with many classic applications of Calculus, especially to Physics. This book also has many exercises throughout it, with the answers to every one of them at the end of the book. This book will teach you the core of first-year Calculus, that is, the two main parts which form the foundation for all of Calculus, the Differential Calculus and the Integral Calculus, and the relationship between these two parts. In addition, I have included a section of the book which will introduce you to the subject of Differential Equations, which is very important in many of the applications of Calculus. I have made the effort to make this level of the Calculus as clear as possible. However, if the student finds some part of the book difficult, I urge the student to at least know the concept and the conclusion of that section, and to keep moving with the text and persevere. Usually, a difficult problem here or there that the student may not understand in a math book, is not a major hindrance to getting benefit from the book.

We have inherited a world with many problems. There are many challenges for us as a species if we are to have any future. I urge the reader to not shrink from mathematics, but rather embrace its challenges. As the late American President John F. Kennedy said about sending men to the moon - "We choose to go to the moon in this decade and do the other things, not because they are easy, but because they are hard."* Mathematics is our greatest tool and our common language for understanding the world, created by man to be understood by man. It is certainly not beyond your ability, and not to be feared. Perhaps you too will someday add to mathematics, solve emerging problems with it, and find new ways that it can be applied.

There is some material concerning numbers and set theory in the first few chapters that is not usually discussed in a Calculus book, but that I think the student may find interesting and thought provoking. It is my sincere belief that the prepared and serious student, willing to faithfully follow along with the reading and the exercises, will find that journey a very rewarding one!

Timothy C. Kearns
BS, Virginia Tech (cum laude), 1983.
October 1, 2015
Fairfax, Virginia. USA

* www.goodreads.com/quotes/22043

Introduction

Calculus is the mathematics of change, and change is an integral part of the universe. Mathematicians and scientists of all persuasions know that calculus is a cornerstone of modern science. Calculus allows us to solve a variety of problems dealing with continuously varying quantities. This development which dates back to the 17th century, with the work of many great mathematicians, but in particular Isaac Newton and Gottfried Leibniz, has added tremendously to the power of our science and has allowed us to understand and master our world in ways that are nothing less than revolutionary. We should consider it to be one of the few truly great achievements of the human mind. Most of the scientific and technological advances of the past few centuries can be ultimately attributed to the development of calculus and related mathematical advances.

This book explains all of the basic concepts of single variable calculus through the theory and application of the derivative, the theory and application of the definite integral, and the connection between these two main parts of the subject, by way of the Fundamental Theorem of Calculus. After the discussion of differentiation and integration, I have included some of the basics of differential equations and their applications so that the student can see how important the differential and integral calculus is to many different areas. The book contains an abundance of examples at every step and many exercises to help the student learn the subject. It has been titled Introductory Calculus because it is mainly about the single variable part of the subject, the portion devoted to real valued functions of a single variable, which is the starting point for most of the larger treatment of calculus. In this way we have a compact and rigorous introduction to calculus, so that the student can quickly grasp the essential concepts and get a feel for the many applications of the subject. This book should help any serious student that needs to know calculus.

Part I

Numbers,
Sequences & Series,
Limits & Continuity

(1) The Real Numbers

In calculus, we are mainly concerned with the real numbers. So students of calculus should have a good understanding of the real numbers and the various important subsets of them.

(#) Integers and Rationals: The usual starting point is the set of positive integers {1, 2, 3, . . . }, also called the natural numbers and given the name N. When we include 0 with N we get the whole numbers W = {0, 1, 2, 3, . . . }. Then combining the negatives of the natural numbers with W gives us the very important set of numbers known as the integers Z: Z = { . . . ,-3, -2, -1, 0, 1, 2, 3, . . . }.

Many times in mathematics it is important to emphasize that some quantity only takes values from N or from Z, and so on. When we consider all numbers of the form $\frac{a}{b}$, where a and b are integers, b ≠ 0, we have the set of rational numbers Q. All rational numbers have decimal expansions that are finite or have an infinite decimal expansion that repeats some pattern forever. It is easy to see why the decimal expansion of a rational number would always eventually repeat some pattern. As an example, consider the rational number $\frac{17}{9}$. Written as a decimal number (using long division), $\frac{17}{9}$ can only have the remainders {0,1,2,3,4,5,6,7,8}. So there will be some repeating pattern, because the decimal expansion ultimately will return to one of these remainders, and cycle through some pattern indefinitely.

(#) Irrational and Real Numbers: The rationals mark the end of what we might call "ordinary" numbers. We are familiar with and accustomed to thinking in terms of whole number or fractional quantities, such as 200 chairs, 10 tables, $\frac{1}{2}$ pound of cheese, $\frac{3}{4}$ of an acre, or $\frac{1}{3}$ of a pumpkin pie, and so on.

Our minds do not as easily comprehend the great abundance of all the other numbers on the real line, the so-called irrational numbers. Irrational numbers have decimal expansions that go on forever without ever repeating any pattern (we can't do long division of irrational numbers). Irrational numbers include numbers like π, $\sqrt[5]{17}$, $\sqrt{2}$, and so on. The rationals and the irrational numbers obviously have no overlap. With the number systems discussed above we had a progression of sorts. We have $N \subset W \subset Z \subset Q$. This progression ends with the rationals. The real numbers consist simply of the union of the two disjoint sets known as the rationals and the irrationals, and we can denote them with the symbol R.

(#) Infinity: All of these number systems are infinite sets, and mathematicians have discovered that there are different types of infinity. In this section, we will address a little bit of the theory of infinite sets, and the ideas of countable and uncountable infinity. While important, the reader may choose to forgo this discussion and also the next section on the measure of numbers, and move to the section about complex numbers because a discussion about infinity and measure is not too critical to the study of calculus. However, the reader may gain a little more insight into the real numbers by reading this section, and some readers may find it rather interesting. The discussion is really not very difficult or excessive.

An infinite set of numbers that can be put into one-to-one correspondence with the positive integers (also of course known as the natural numbers N) is said to be countably infinite. The natural numbers N are then themselves of course countably infinite, and this is considered to be the lowest or simplest type of infinity. Consider the set of all positive multiples of 20, written A = {20, 40, 60, . . .} or the set of all the positive even numbers B = {2, 4, 6, . . .}. We can set up a one-to-one correspondence between these two sets A and B with N by way of the functions
f(n) = 20n, n = 1,2,3, . . ., and g(n) = 2n, n = 1,2,3, . . ., respectively. Clearly the sets A and B have the same number of elements as N because of these one-to-one correspondences, and we then say that the three sets

A, B, and N are equivalent. It is clear that there are an infinite number of countably infinite sets. The sets A and B are subsets of N, but have the same number of elements. This is a counter-intuitive property for sets to have. It is not true with finite sets, but it is true with any infinite set. It leads mathematicians to define an infinite set as a set which is equivalent to a subset of itself. In set theory, we call the number of elements in a set its cardinality. There is no finite number which we can use as the cardinality of an infinite set, so we have to have a symbol known as a transfinite number to denote the type of infinity associated with an infinite set. Mathematicians use the symbol $\aleph_0$ to denote the cardinality of countably infinite sets (this is pronounced Aleph-zero, using the first letter of the Hebrew alphabet).

To show that the integers Z are countably infinite we can list the elements of Z as {0, 1, -1, 2, -2, 3, -3, . . . }. Then consider the function h(n) defined in the following way: h(n) = { (1,0), (2,1), (3,-1), (4,2), (5,-2), . . . }, where n is the first number in each ordered pair. This function defines a one-to-one correspondence between N and Z. Therefore the integers Z are countably infinite with cardinality $\aleph_0$.

Now we wish to show that the set of rational numbers Q is also countably infinite. That this is true of Q goes against our intuition, since there are an infinite number of rationals between any two of them. To show that Q is countably infinite, consider the x-y plane, and in particular only the points (x,y) in the plane where x and y are integers. Associate with every point (x,y) in this lattice the rational number $(\frac{x}{y})$. Let's throw out every point on the x-axis except the origin, because we can't divide by 0. Also throw out every point on the y-axis except the origin, because they would all correspond to zero. We haven't thrown out the origin (0,0), because we want to associate this point with the rational number 0. Now, follow the path through this lattice as shown in the illustration below. Among all the other points in the lattice, there will be many repeated numbers. For example, (1,2) and (8,16) and many other points all correspond to the fraction $\frac{1}{2}$. After throwing out repeats, after throwing out all numbers on the x-axis other than (0,0), and throwing out all points on the y-axis other than (0,0), we will then be able to one by one list all of the rational numbers. Much as was done above with the integers Z, the infinite list $\{0,-1,1,2,-2,-\frac{1}{2},\frac{1}{2},\frac{3}{2},3,-3,-\frac{3}{2},-\frac{2}{3},-\frac{1}{3},\frac{1}{3},\frac{2}{3},\ ...\}$ obtained by following the path in the illustration shows that the rationals Q are countably infinite, and therefore have cardinality $\aleph_0$.

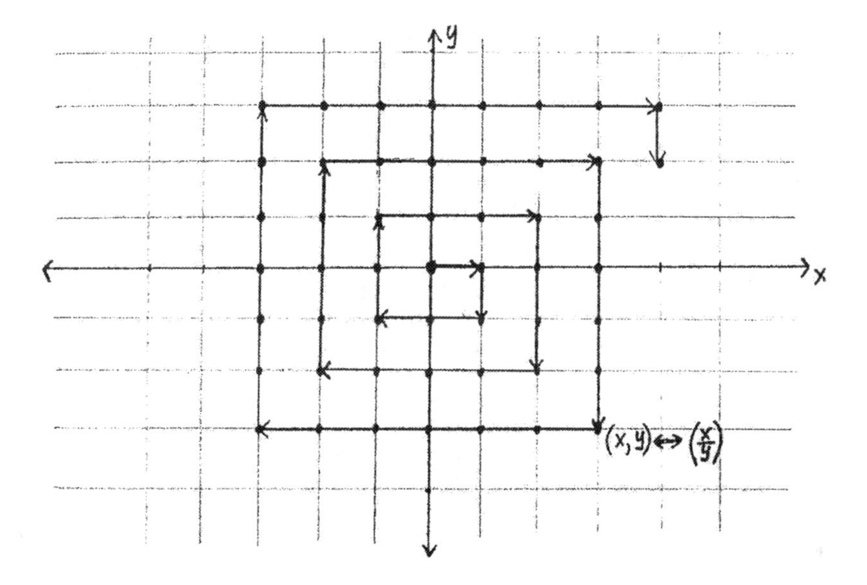

Now we want to show that the irrationals and hence the real numbers are uncountably infinite, that is, the infinity of them is a larger type of infinity than sets with cardinal number $\aleph_0$. We say that an uncountably infinite set of numbers has cardinality $\aleph_1$. We will just consider the set of all real numbers in the interval [0,1]. Let's assume that the real numbers in [0,1] are countably infinite. Then we should be able to list all of them. The least number in the list is (0.0000 . . .) and the greatest number is (0.9999 . . .). The number (0.9999 . . .) is actually another way of writing 1 and the reader should be able to show this once we study geometric series in the next chapter. Let the listing of rationals and irrationals in [0,1] have the following form, where all subscripted letters are our familiar decimal digits from the set {0,1,2,3,4,5,6,7,8,9}:

$$0.a_1\ a_2\ a_3\ a_4\ ...$$

$0.b_1\ b_2\ b_3\ b_4\ \ldots$
$0.c_1\ c_2\ c_3\ c_4\ \ldots$
$0.d_1\ d_2\ d_3\ d_4\ \ldots$

. .

. .

. .

Now construct a number z = $(0.z_1\ z_2\ z_3\ z_4\ \ldots)$, where $z_1 \neq a_1$, $z_2 \neq b_2$, $z_3 \neq c_3$, $z_4 \neq d_4$, and so on. Then the number z will differ from the first number in the list in at least the first decimal place, it will differ from the second number in the list in at least the second decimal place, it will differ from the third number in the list in at least the third decimal place, and so on. Therefore the number z is not on the list. This is a contradiction of our assumption that all numbers in [0,1] could be listed because they're countably infinite. We could add z to the list and go through this again with the same result. So mathematicians have deduced that the infinity of real numbers in [0,1] must be a larger type of infinity than countable infinity, and they call it uncountable infinity. They denote its cardinality $\aleph_1$, which is supposedly greater than $\aleph_0$. Mathematicians have tried to find a type of infinity between these two, but have had no success. It is believed that this uncountable infinity of real numbers constitutes a continuum, whereas the countable infinity of rationals does not.

There are ever higher types of infinity, an infinite number of them in fact, but they are important in advanced set theory and not in calculus. These higher types of infinity undoubtedly correspond to points or objects of some kind on ever more complicated geometric sets, which we can't even begin to visualize or describe. In mathematics we do have some limitations to what we can know and do. Calculus deals primarily with what we experience in the world that we apparently live in, a world where we work with things in one, two, and three dimensions.

(#) The Measure of Numbers: In higher mathematics, there is the subject that we call measure theory. The measure of subsets of the real numbers on the real line can be thought of as the collective "width" of the members of the set. It turns out that the measure of countably infinite sets of numbers on the real line, like Z and Q, is zero. For Q, it is easy to prove this with an argument involving open sets of a certain type that cover each of the members of the set, and then demonstrating how the collective width of the cover can be made arbitrarily small while still covering every rational number. This proves that the measure of Q is zero. The proof once again uses geometric series which we study in the next chapter. Since the measure of Q is zero, it is as if they are not even there, from a measure theoretic point of view. Therefore, the set of rationals on the real line do not form a continuum, there remains holes or missing pieces, so to speak. However the set of real numbers, with the rationals and irrationals combined, do form a continuum on the real line, so that the measure of a real number set like [a,b] or (a,b) is as we would expect, (b-a). On the real line, any point corresponds to one real number, and every real number corresponds to only one point. Another property of the real numbers that we should mention is that they are ordered. This means that if x and y are any two real numbers, then either $x < y$, $x = y$, or $x > y$.

(#) Complex Numbers: There is another number set which is of major importance in mathematics. It is the set of complex numbers. These numbers correspond to all the points in a plane (called the complex plane). A complex number is any number of the form (a + bi), where a and b are real numbers and i is the imaginary unit $\sqrt{-1}$. Thinking in terms of coordinates, (a,b) would correspond to the complex number (a + bi). The number a is the real part and bi is the imaginary part. The real numbers correspond to the horizontal axis in the complex plane. They are the complex numbers of the form (a + 0i), where there is no imaginary component. So we can understand why the reals are ordered from $(-\infty)$ on the left up to $(+\infty)$ on the right. The reals are numbers on a line, increasing from left to right. All other numbers in the complex plane have a real component a (a could be 0) and an imaginary component b ($b \neq 0$). When we think about two numbers in the complex plane, we can understand why complex numbers are not considered to be ordered (how can all the points in a plane be ordered like the points on a line are?). We will not deal with complex numbers too much in this book about calculus. Calculus is mainly concerned with real-valued functions of a real variable, (or variables) and making calculations with these types of functions.

(2) Sequences and Series

Sequences are very important in calculus. In your study of this subject, the observant student will notice just how pervasive they are. We will always find ourselves talking of processes that continue forward in a sequential fashion toward some limiting state. I must confess that this discussion of sequences is a bit theoretical and probably a bit too much for most students at this stage. So those who find this section a bit too theoretical should merely concentrate on the basic definitions and concepts presented in the early part of the section and not worry about the rest.
The well prepared students that easily follow the discussion should benefit from reading this section. There will be no exercises for this section. In the section after this we will discuss the topic of limits, which is very important in calculus. To speak of the limit of some quantity, we will depend on the notion of a sequence.

(#) Sequences: A sequence is a function whose domain is the set of positive integers. Abstractly, a sequence is written:

$$\{ a_1, a_2, a_3, \ldots \} = \{a_n\}$$

We can think of this sequence, since we said it was a function, as $\{y(1), y(2), y(3), \ldots\} = \{y(n)\}$, where $y(n) = a_n$. Obviously the domain is the set N = {1,2,3, . . . }, which is the set of positive integers, which are also known as the natural numbers N. So in the above description we are listing only the values of the dependent variable. A sequence consists of a countably infinite set of points. Contrast this type of function with the more commonly encountered functions in calculus, like for example $y = y(x) = x^2 + 3x + 9$. This is a parabola, which is a continuous function of the real variable x. We say that the domain of y(x) is the set of real numbers, an uncountably infinite set. A sequence is a type of set, where each element, a function value, may occur more than once (because we know that for a function the same value can be assigned to more than one member of the domain). Also for sequences, unlike more general sets where the order that we write down the elements of the set is unimportant, the order in which we write the elements of a sequence is important and has meaning. The following are examples of sequences:

$\{1, \frac{1}{2}, \frac{1}{3}, \ldots\} = \{\frac{1}{n}\}$, n = 1,2,3, . . .

$\{2, 3, 4, \ldots\} = \{n + 1\}$, n = 1,2,3, . . .

$\{\frac{1}{3}, \frac{8}{4}, \frac{27}{5}, \ldots\} = \{\frac{n^3}{n+2}\}$, n = 1,2,3, . . .

$\{ 0, 2, 0, 2, \ldots\} = \{ 1 + (-1)^n \}$, n = 1, 2, 3, . . .

(#) <u>Convergence and Divergence</u>: We need to discuss the topic of convergence or divergence of a sequence, which is the most important consideration for us in our discussion of sequences. A sequence is said to converge if there is a number L that the terms of the sequence approach arbitrarily closely, for all terms after a certain point. For example, the sequence $\{ 1, \frac{1}{4}, \frac{1}{9}, \frac{1}{16}, ... \} = \{\frac{1}{n^2}\}$ clearly converges to zero. The sequence $\{ 0, 1, 0, 1, ... \}$ diverges because there is no single number L that the terms are getting closer and closer to as $n \to \infty$. The exact definition of convergence for a sequence $\{a_n\}$ is:

The sequence $\{a_n\}$ converges to a number L if for any arbitrarily small positive number ε, there exists a positive number M such that $|a_n - L| < \varepsilon$ whenever $n > M$.

There is another equivalent definition of convergence that is called Cauchy convergence which mathematicians often find useful:

The sequence $\{a_n\}$ converges to a number L if for any arbitrarily small positive number ε, there exists a positive number M such that $|a_n - a_m| < \varepsilon$, whenever n and m are both greater than M.

Example 1: Consider the sequence given above: $\{ 1, \frac{1}{4}, \frac{1}{9}, \frac{1}{16}, ... \} = \{\frac{1}{n^2}\}$ and let $\varepsilon = \frac{1}{100}$. Then for M = 10, $|\frac{1}{n^2} - 0| < \varepsilon$, whenever $n > M$. Since we could have chosen ε as small as we like, it is clear that $\{\frac{1}{n^2}\}$ converges to 0, as $n \to \infty$.

Example 2: Consider the sequence $\{0, \frac{1}{2}, \frac{2}{3}, \frac{3}{4}, ... \} = \{\frac{n-1}{n}\} = \{1 - \frac{1}{n}\}$, and let $\varepsilon = \frac{1}{1000}$. Then for M = 1000, $|(\frac{n-1}{n}) - 1| = |(1 - \frac{1}{n}) - 1| = |-\frac{1}{n}| = \frac{1}{n} < \varepsilon$, whenever $n > M$. Since we could have chosen ε as small as we like, it is clear that $\{\frac{n-1}{n}\}$ converges to 1 as $n \to \infty$.

Example 3: Consider the sequence $\{0, 1, 0, 1, ...\}$. For $0 < \varepsilon < \frac{1}{2}$,

$|a_n - a_{n+1}| > \varepsilon$ for all n = 1, 2, 3, . . . This says that any two consecutive terms always differ by more than ε; so the terms are not approaching any single number L arbitrarily close as $n \to \infty$. So this sequence is divergent. For this sequence, there are two subsequences (an infinite subset of the terms of the sequence) {0, 0, 0, . . . } and {1, 1, 1, . . . } that do converge to 0 and 1 respectively. The first of these subsequences consists of all the odd numbered terms, the second subsequence consists of all the even numbered terms. It should be noted that there are an infinite number of subsequences for any given sequence, and that a sequence converges to a number L iff (if and only if) all of its subsequences converge to L also.

Example 4: Consider the sequence {1, 2, 3, . . . } = {n}. Obviously, the terms are tending to ∞, as $n \to \infty$. So this sequence diverges.

We say that a set of numbers is bounded above if there is a number B such that $x \leq B$ for all members x of the set. We say that a set is bounded below if there is a number B such that $x \geq B$ for all members x of the set. For upper bounds and for lower bounds, we have the concept of a least upper bound and of a greatest lower bound. A set S has a least upper bound α, lub(S) or supremum(S), written sup(S), if α is an upper bound, and if for any other upper bound β, we have $\alpha \leq \beta$. Similarly, a set S has a greatest lower bound α, glb(S) or infimum(S), written inf(S), if α is a lower bound, and if for any other lower bound β, we have $\alpha \geq \beta$.

Example 5: Consider the sequence $\{ \frac{sin(1)}{1}, \frac{sin(2)}{2}, \frac{sin(3)}{3}, . . . \} = \{ \frac{sin(n)}{n} \}$. Let's choose $\varepsilon = \frac{1}{300}$. So now, we have to determine if there is a number L such that $|\frac{sin(n)}{n} - L| < \varepsilon$ for all terms after a certain point. Note that sin(x) is bounded below and above, because we know $-1 \leq \sin(x) \leq 1$ for any real number x. Also $|\frac{sin(n)}{n}| = \frac{|sin(n)|}{|n|} \leq \frac{1}{n}$ for $n \geq 1$, since $|\sin(n)| \leq 1$ for all n. So it appears that L = 0 is what the sequence converges to. Let's let M = 300. Then $|a_n - L| = |\frac{sin(n)}{n} - 0|$ $\leq \frac{1}{n} < \frac{1}{300}$ whenever n > M. Since we could have chosen ε as small as we like, and we would always be able to find an M corresponding to our choice of ε, we can see that the sequence converges to 0.

(#) <u>Bounded Monotone Sequences</u>: In our discussion of sequences, there are two more important points that we should consider:

If we have a sequence $\{a_n\}$ that is monotonically increasing and bounded above, then the sequence will converge to some number L. Similarly, if we have a sequence $\{a_n\}$ that is monotonically decreasing and bounded below, then the sequence will converge to some number L. Let's prove the first assertion, the second assertion follows similarly. We are given that the sequence $\{a_n\}$ is bounded above, so $\sup\{a_n\} = \alpha$ exists. Let $\varepsilon > 0$. Obviously there is a member of the sequence in the interval $[\alpha - \varepsilon, \alpha]$ and since the sequence is always increasing there must be an infinite number of terms in this interval. Then it is clear that $|\alpha - a_n| < \varepsilon$ for all terms a_n after some point. Since ε can be arbitrarily small, we have shown that the sequence converges to α.

<u>Example 1</u>: The sequence $\{2 - \frac{1}{\sqrt{n}}\}$ is always less than or equal to 2, so it is bounded above by 2. Since $\left(2 - \frac{1}{\sqrt{n+1}}\right) > \left(2 - \frac{1}{\sqrt{n}}\right)$ for $n \geq 1$, the sequence $\{2 - \frac{1}{\sqrt{n}}\}$ is monotonically increasing. Therefore it will converge to some number L. We can see that the sequence converges to L = 2.

<u>Example 2</u>: The sequence $\{1 + \frac{cos^2(n)}{2n}\}$ is always greater than or equal to 1, so it is bounded below by 1. Also, we can see that $\left(1 + \frac{cos^2(n+1)}{2(n+1)}\right) <$ $\left(1 + \frac{cos^2(n)}{2n}\right)$ for $n \geq 1$. So the sequence $\{1 + \frac{cos^2(n)}{2n}\}$ is monotonically decreasing. Therefore this sequence will converge to some number L. We can see that the sequence converges to L = 1.

(#) **Infinite Series**: This is a rather extensive topic in calculus, that is not usually discussed before differentiation and integration, so we will not dwell on it too much here. The reader may find it instructive to talk of it because of its connection with sequences, and it will not harm the reader to learn some of the basics of infinite series.

An infinite series is an infinite sum, written $\sum_{n=1}^{\infty} a_n = (a_1 + a_2 + a_3 + \cdots)$.

The individual terms of an infinite series may be 0, negative, or positive. The series often starts with n = 1, or with n = 0, or any other integer, and this can be important depending on the application. The convergence or divergence of the series is one of our main considerations, as it is with sequences. By convergence, we mean that the sum approaches arbitrarily closely some finite number L as $n \to \infty$. If there is no finite number L that the infinite series approaches arbitrarily closely as $n \to \infty$, then we say that the series diverges.

The correct way to consider the convergence or divergence of an infinite series is by considering the so-called sequence of partial sums $\{S_n\}$, which we will now discuss.

If we have the series $\sum_{n=1}^{\infty} a_n$, we define the terms of the sequence of partial sums in the following way:

$S_1 = a_1$,
$S_2 = a_1 + a_2$,
$S_3 = a_1 + a_2 + a_3$, and so on to ∞.

If the sequence of partial sums $\{S_n\}$ converges to a finite number L, then we say that the infinite series converges to L (adds up to L). If the sequence of partial sums $\{S_n\}$ diverges to $(-\infty)$ or $(+\infty)$, then the corresponding infinite series is divergent (adds up to $(-\infty)$ or $(+\infty)$).

The sequence of partial sums could also oscillate back and forth, not getting arbitrarily close to any number L as $n \to \infty$, in which case we would also say that the corresponding infinite series diverges.

(**Geometric Series**)

There are several main types of infinite series, but we are only prepared at this point to introduce one rather important one, geometric series. Let a be any real number, and let r be any real number with |r| < 1. Then

$\sum_{n=0}^{\infty} a(r)^n = (a + ar + ar^2 + ar^3 + \cdots) = a(1 + r + r^2 + r^3 + \cdots)$

$= a \sum_{n=0}^{\infty} (r)^n$ is called a geometric series, and r is the common ratio.

To find the sum of this series, consider the first (n + 1) terms multiplied by $(1 - r)$:

$(1 - r)(a) \sum_{k=0}^{n} (r)^k$ = (a) $[(1 + r + r^2 + r^3 + \cdots + r^n) - (r + r^2 + r^3 + \cdots + r^{n+1})]$

= (a) $(1 - r^{n+1})$.

Therefore, $\sum_{k=0}^{n} (a)(r)^k = (a)\frac{(1 - r^{(n+1)})}{(1 - r)}$.

Since |r| < 1, $(r^{n+1}) \to 0$ as $n \to \infty$, and

the sum of a geometric series is $\sum_{n=0}^{\infty} (a)(r)^n = \frac{(a)}{(1 - r)}$. Note that this result depends on n starting at n = 0.

Example 1: Use a geometric series to show that (0.535353 . . .) = $\frac{53}{99}$.

(0.535353 . . .) = $\sum_{n=1}^{\infty} (53)(\frac{1}{100})^n = [\sum_{n=0}^{\infty} (53)(\frac{1}{100})^n - 53(\frac{1}{100})^0] = (\frac{53}{1 - \frac{1}{100}} - 53)$

(The sum of a geometric series starting with n = 1 is the sum of the geometric series starting with n = 0 minus the term corresponding to 0)

$= ((\frac{5300}{99}) - 53) = (\frac{5300 - 5247}{99}) = \frac{53}{99}$.

Example 2: The infinite sum $\sum_{n=0}^{\infty}(13)(\frac{1}{7})^n = \frac{13}{(1-\frac{1}{7})} = \frac{13}{(\frac{6}{7})} = \frac{91}{6} \approx 15.166$,

since it is a geometric series with a = 13, r = $\frac{1}{7}$.

Example 3: We will digress for a moment to a topic in the chapter on real numbers, concerning the measure of the rationals Q. If you skipped that discussion, then you can skip this example. For the set of rationals Q, which is a countably infinite set and can therefore in principle be listed because of their one-to-one correspondence with the natural numbers, we will show that they can be covered with a collection of open sets, the collective width of which can be made arbitrarily small (that is, the measure of the rationals or any countably infinite set for that matter, is zero).

Proof: Let $\varepsilon > 0$ be an arbitrarily small positive number. Since the set of rationals Q is countably infinite, they can be listed: $\{q_1, q_2, q_3, \ldots\}$. Cover the number q_i with the open set $(q_i - \frac{(\varepsilon)^i}{2}, q_i + \frac{(\varepsilon)^i}{2})$, (i = 1,2, . . .). The collective width of these covering sets is $(\varepsilon + \varepsilon^2 + \varepsilon^3 + \cdots)$, which is a geometric series which converges when $|\varepsilon| < 1$ and equals $(\frac{1}{1-\varepsilon} - 1) = (\frac{\varepsilon}{1-\varepsilon})$. This converges to 0, as $\varepsilon \to 0$. So we say that the measure of the rational numbers on the real line is 0, in other words it is like they are not even there at all.

(3) Limits and Continuity

The notion of a limit is very important in calculus. The main thing that we want to consider is the limiting value of some function y(x), as the independent variable x approaches some fixed number "a" from the left, from the right, or from both sides simultaneously (without ever reaching a). These limits are written respectively as:

$$\lim_{x \to a^-} y(x) \quad , \quad \lim_{x \to a^+} y(x) \quad , \quad \text{and} \quad \lim_{x \to a} y(x)$$

The first two are called one-sided limits, and the last one is the two-sided limit, or just the limit. For the purposes of illustration, let's assume a = 5.

(#) <u>Left-Sided Limits</u>: For the case $x \to a^-$. Since a = 5, then we consider some sequence like {4.9, 4.99, 4.999, . . .}. This is a sequence where the terms are approaching 5 from the left (the negative side), but they never reach 5 . Any such sequence where the terms are approaching 5 from the left is what we want to consider, and there are an infinite number of them. For example, a sequence like the following: {4.8, 4.934, 4.979, . . . } would be sufficient also, as long as it is understood that the terms are approaching 5 from the left arbitrarily closely, without ever reaching 5.

So for the above one-sided limit $\lim_{x \to 5^-} y(x)$ we are interested in the convergence or divergence of the sequence of function values: {y(4.9), y(4.99), y(4.999), . . . }

(#) <u>Right-Sided Limits</u>: For the case $x \to a^+$, using a = 5, we consider some sequence like {5.1, 5.01, 5.001, . . . }. This is a sequence whose terms approach 5 arbitrarily closely from the right (the positive side) without ever reaching 5. Once again, as in the case of left-sided limits there are an infinite number of such sequences.

So for the one-sided limit $\lim_{x \to 5^+} y(x)$ we are interested in the convergence or divergence of the sequence of function values:
{y(5.1), y(5.01), y(5.001), . . . }

(#) <u>The Limit</u>: For the case $x \to a$, again assume that a = 5. This is called a two-sided limit, or just the limit. We can consider some sequence of x-values like {5.1, 4.9, 5.01, 4.99, 5.001, 4.999, . . . } where we approach 5 from both sides arbitrarily closely without ever reaching 5. Once again, there are an infinite number of such sequences. So we are interested in the convergence or divergence of the corresponding sequence of function values:
{y(5.1), y(4.9), y(5.01), y(4.99), y(5.001), y(4.999), . . . }.

It turns out that the limit $\lim_{x \to a} y(x)$ exists iff $\lim_{x \to a^-} y(x) = \lim_{x \to a^+} y(x)$.
The sequence of function values just considered contains two main subsequences: one corresponding to all of the terms where x was to the left of a, and one corresponding to all of the terms where x was to the right of a. Since we know that a sequence converges to a single number iff all of its subsequences converge to the same number, we can then see why we require that the two one-sided limits exist and be equal to each other for the limit to exist.

(#) <u>Properties of Limits</u>: Suppose we have two functions y = f(x) and y = g(x), a is a real number, and n is rational. Furthermore, assume

$\lim_{x \to a} f(x) = L$ and $\lim_{x \to a} g(x) = M$, where L and M are both finite.

Then, $\lim_{x \to a} (f(x) \pm g(x)) = \lim_{x \to a} f(x) \pm \lim_{x \to a} g(x) = (L \pm M)$.

$\lim_{x \to a} f(x) \cdot g(x) = \lim_{x \to a} f(x) \cdot \lim_{x \to a} g(x) = (L \cdot M)$.

$\lim_{x \to a} \frac{f(x)}{g(x)} = \frac{\lim_{x \to a} f(x)}{\lim_{x \to a} g(x)} = \left(\frac{L}{M}\right)$, $M \neq 0$.

$\lim_{x \to a} (f(x))^n = (L)^n$

(#) <u>Indeterminate Forms</u>: When considering limits, many times we encounter various types of indeterminate forms, where we cannot directly determine what the limit is. In these cases, we have to algebraically change the quantity to a form where we can figure out the limit. The types of indeterminate forms that we encounter most are $\left(\frac{0}{0}\right)$, $\left(\frac{\infty}{\infty}\right)$, $(0 \cdot \infty)$, (1^{∞}), and $(\infty - \infty)$.

<u>Example 1</u>: Find the $\lim_{x \to 7} 2x^2 - 3x + 1$.

$\lim_{x \to 7} 2x^2 - 3x + 1 = 2(7)^2 - 3(7) + 1 = 78.$

<u>Example 2</u>: Find the $\lim_{x \to 1} \frac{x^2 + 4}{3x}$.

$$\lim_{x \to 1} \frac{x^2 + 4}{3x} = \frac{(1)^2 + 4}{3(1)} = \frac{5}{3}.$$

<u>Example 3</u>: Find the $\lim_{x \to 2^+} \frac{3}{x^2 - 4}$.

For numbers approaching 2 from the right, the quantity $(x^2 - 4)$ is approaching 0, but $(x^2 - 4)$ is always positive. So $\frac{3}{x^2 - 4}$ is tending to (3 divided by a positive quantity that approaches 0), therefore the limit is $(+\infty)$.

<u>Example 4</u>: Find the $\lim_{x \to 6} \frac{(x - 6)}{x^2 - 5x - 6}$.

When we substitute 6 for x, we get the indeterminate form $\left(\frac{0}{0}\right)$, which means that we don't know what the fraction is approaching. So we must algebraically manipulate this quantity in some way to get the limit.

So, the $\lim_{x \to 6} \frac{(x-6)}{x^2 - 5x - 6} = \lim_{x \to 6} \frac{(x-6)}{(x-6)(x+1)} = \lim_{x \to 6} \frac{1}{(x+1)} = \frac{1}{7}$.

Example 5: Find the $\lim_{x \to \infty} \frac{3x^3 + 4x + 1}{2x^3 - 5x^2 + x + 2}$.

As x $\to \infty$, both numerator and denominator go to ∞, so we have the indeterminate form $\frac{\infty}{\infty}$, which means that we don't know what the fraction is approaching. However, if we divide every term in the numerator and denominator by x^3, we get,

$\lim_{x \to \infty} \frac{3x^3 + 4x + 1}{2x^3 - 5x^2 + x + 2} = \lim_{x \to \infty} \frac{(3 + \frac{4}{x^2} + \frac{1}{x^3})}{(2 - \frac{5}{x} + \frac{1}{x^2} + \frac{2}{x^3})}$. As x $\to \infty$, every term in the top and the bottom that is a fraction will tend to 0. Therefore the limit is $\left(\frac{3+0+0}{2-0+0+0}\right) = \frac{3}{2}$.

(#) Continuity: When we have a function y(x) defined on an interval [a,b] and c $\in$ (a,b), then we say that the function y(x) is continuous on the open interval (a,b) at x = c, if the $\lim_{x \to c} y(x) = y(c)$, for all c $\in$ (a,b). Note that this says three things: the limit must exist at c, the function must be defined at c, and the two must be the same.

In addition, if $\lim_{x \to a^+} y(x)$ = y(a), the function y(x) is right continuous at a,

and if $\lim_{x \to b^-} y(x)$ = y(b), the function y(x) is left continuous at b.

Then we could say that y(x) is continuous on the closed interval [a,b]. The following figures show ways that a function can be discontinuous.

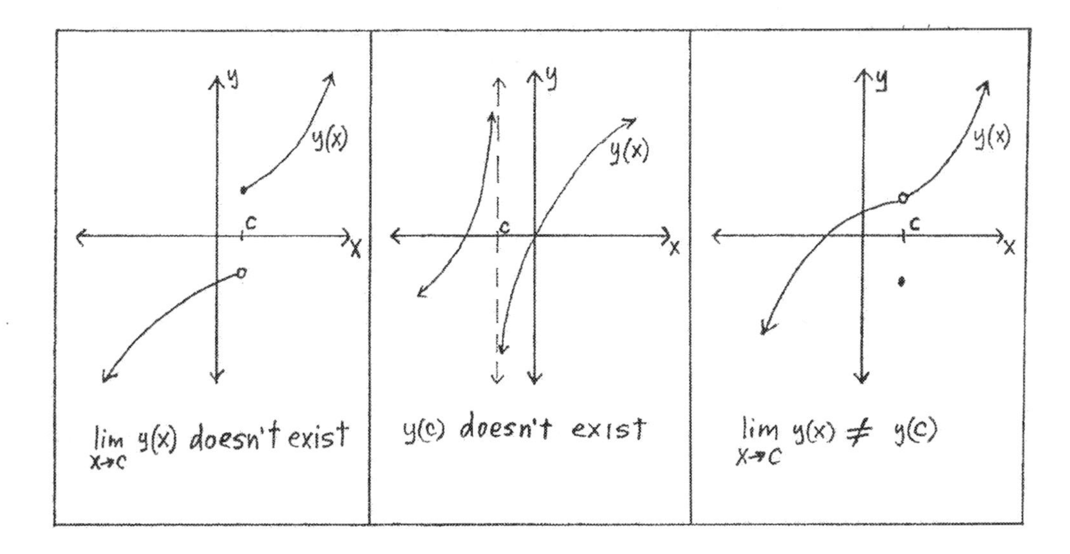

(#) <u>Common Functions</u>: From pre-calculus, the student should be familiar with a lot of the functions most commonly encountered in calculus. Let's briefly review what these functions are, where they are defined and continuous, and the values that they take. After this discussion we will present graphs of many of these important functions.

All polynomials (lines, parabolas, cubics, etc.) are defined and continuous for all real numbers. The range varies with the function, but generally lines, cubics, and odd degree polynomials will range over all real numbers. Even degree polynomials like parabolas, fourth degree polynomials, and so on, will take all values less than or equal to some number or greater than or equal to some number. Rational functions, which are the ratio of two polynomials, are defined and continuous for all real numbers except for those x-values that make the denominator zero.

All even root functions like y = $\sqrt{x}$, y = $\sqrt[4]{x}$, and so on, are defined and continuous when x is greater than or equal to 0. The range of such functions is all real numbers greater than or equal to 0.

All odd root functions like y = $\sqrt[3]{x}$, y = $\sqrt[5]{x}$, and so on, are defined and continuous for all real numbers. The range of such functions is also all of the real numbers.

The exponential functions like y = e^x, y = 2^{-x}, and so on, are defined and continuous for all real numbers. The range of these functions is all positive real numbers. The logarithmic functions like y = ln(x), y = $log_{10}(x)$ are defined and continuous for all positive real numbers. The range of these functions is all real numbers.

The trigonometric functions y = sin(x) and y = cos(x) are defined, periodic, and continuous for all real numbers, with no vertical asymptotes or other breaks. The range of sin(x) and cos(x) is all real numbers in the interval [-1,1]. These two functions are the familiar sinusoidal functions, and they are out of phase with each other by $\frac{\pi}{2}$ radians (90 degrees). The function y = tan(x) = $\frac{sin(x)}{cos(x)}$ is defined, periodic, and continuous for all real numbers except where cos(x) is 0, that is at x = $\{\frac{\pi}{2} \pm k\pi \mid k \in Z\}$, with vertical asymptotes at these x-values. The range of tan(x) is all real numbers. The function y = cot(x) = $\frac{cos(x)}{sin(x)}$ is defined, periodic and continuous for all real numbers except where sin(x) is 0, that is at x = $\{\pi \pm k\pi \mid k \in Z\}$, with vertical asymptotes at these x-values. The range of cot(x) is all real numbers. The functions y = sec(x) = $\frac{1}{cos(x)}$ and y = csc(x) = $\frac{1}{sin(x)}$ are defined, periodic and continuous for all real numbers except where cos(x) and sin(x) are 0 respectively, and these x-values have been described above. There are vertical asymptotes at these x-values. The range of these two functions is all real numbers in $(-\infty, -1] \cup [1, +\infty)$, but they are out of phase with each other by $\frac{\pi}{2}$ radians (90 degrees).

We will now show the graphs of many of the common functions described in this section:

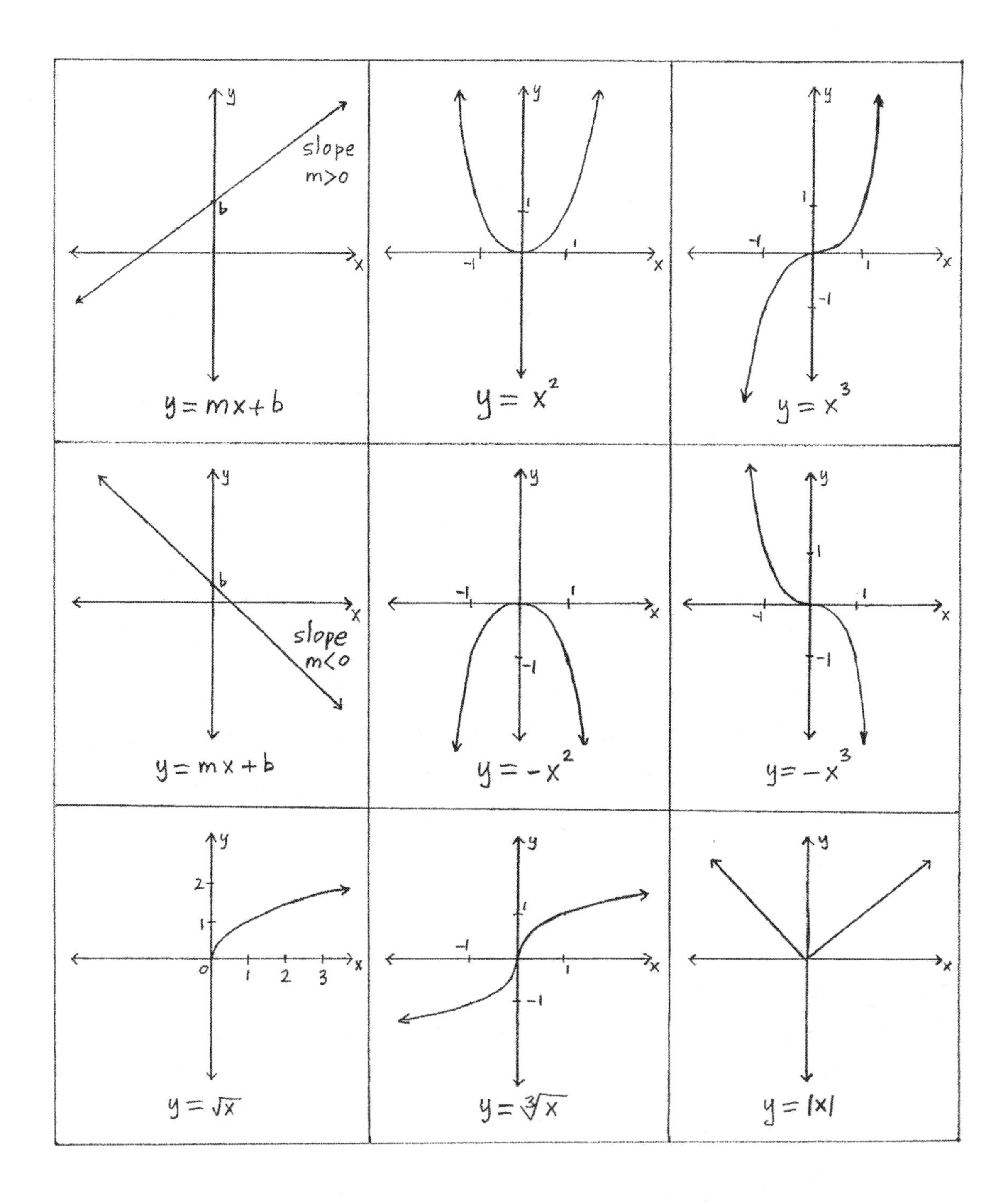

y
slope
m>0
b
x
y = mx + b
y
1
-1
1
x
y = x²
y
1
-1
-1
1
x
y = x³
y
b
x
slope
m<0
y = mx + b
y
-1
1
-1
x
y = -x²
y
1
-1
1
-1
x
y = -x³
y
2
1
0
1
2
3
x
y = √x
y
1
-1
1
-1
x
y = ∛x
y
x
y = |x|

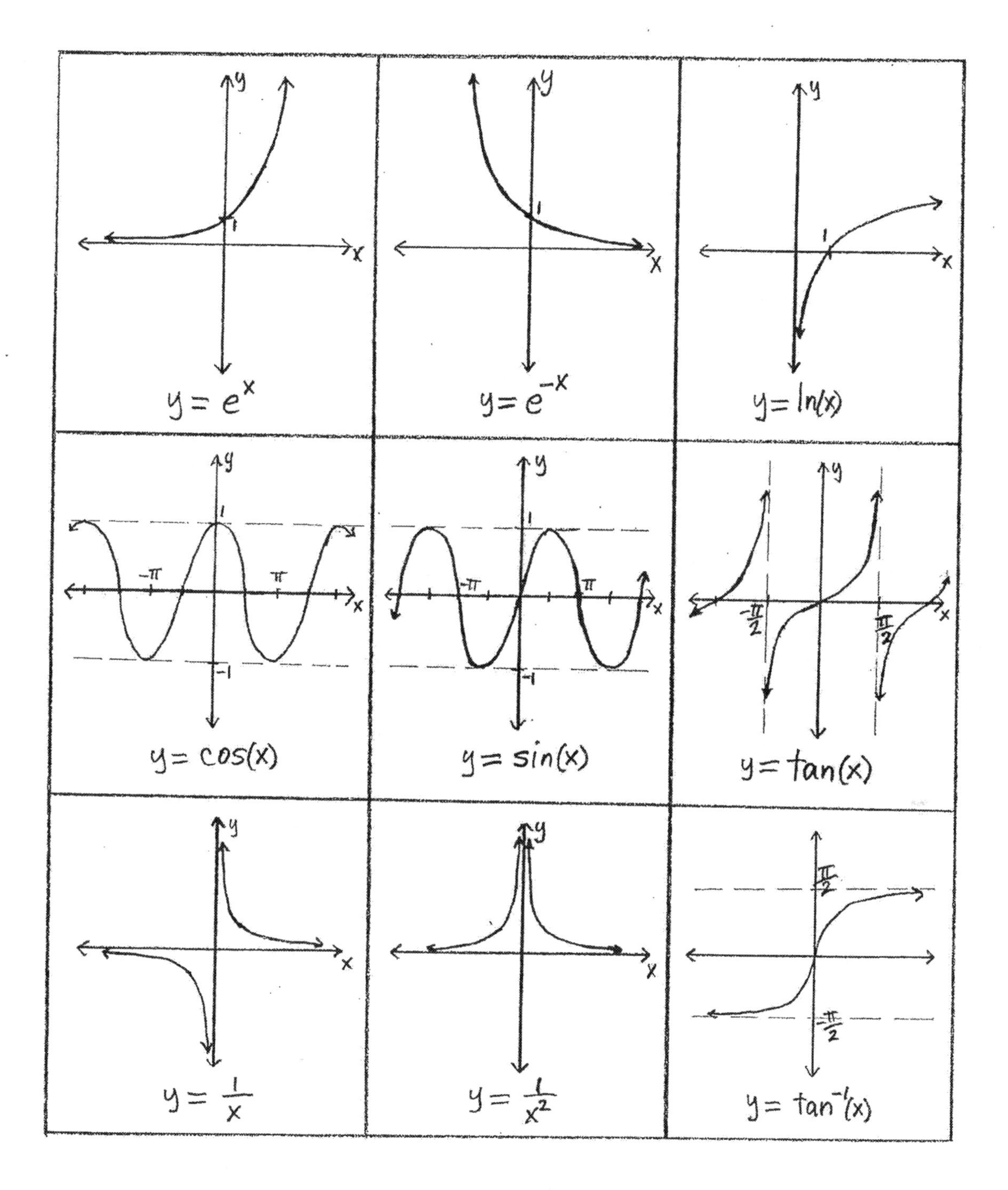
y
x
1
$y = e^x$
y
x
1
$y = e^{-x}$
y
x
1
$y = \ln(x)$
y
x
1
-1
$-\pi$
π
$y = \cos(x)$
y
x
1
-1
$-\pi$
π
$y = \sin(x)$
y
x
$-\frac{\pi}{2}$
$\frac{\pi}{2}$
$y = \tan(x)$
y
x
$y = \frac{1}{x}$
y
x
$y = \frac{1}{x^2}$
$\frac{\pi}{2}$
$-\frac{\pi}{2}$
$y = \tan^{-1}(x)$

EXERCISES:

Evaluate the following limits:

(1) $\lim_{x \to 3} (2x^2 - 9)$

(2) $\lim_{x \to 6} \frac{x-6}{x^2+6}$

(3) $\lim_{x \to 2} \frac{2}{x}$

(4) $\lim_{x \to 0^+} \frac{10}{x}$

(5) $\lim_{x \to 0^-} \frac{10}{x}$

(6) $\lim_{x \to 4} \frac{x-4}{x^2-16}$

(7) $\lim_{x \to 2} \frac{x^2-x-2}{x^2+x-6}$

(8) $\lim_{x \to 4} \frac{x^2-6x+8}{x^2-13x+36}$

(9) $\lim_{x \to 4^-} \frac{1}{x-4}$

(10) $\lim_{x \to 5^+} \frac{1}{5-x}$

(11) $\lim_{x \to 5^-} \frac{1}{5-x}$

(12) $\lim_{x \to 0^-} \frac{x}{x^2}$

(13) $\lim_{x \to -\infty} \frac{x^2+3}{2x^2+1}$

(14) $\lim_{x \to \infty} \frac{2}{x}$

(15) $\lim_{x \to \infty} \frac{x+2}{x}$

(16) $\lim_{x \to -\infty} \frac{x+2}{4x}$

(17) $\lim_{x \to -\infty} \frac{x+2}{4x^2}$ (Hint: divide top and bottom by x^2)

(18) $\lim_{x \to \infty} \frac{2x^2+3x+4}{-3x+9}$ (Hint: divide top and bottom by x^2)

Where are the following functions discontinuous?

(19) y = $\frac{1}{x-1}$

(20) y = $\frac{1}{x^2-1}$

(21) y = $\frac{4x^2}{x^2-2x-8}$

(22) y = $\frac{3x^3}{x^2+1}$

Part II

The Differential Calculus

(4) The Derivative

One of the most important tools that we use in calculus is called the derivative. In this introduction to calculus we are studying the analysis of real-valued functions of a real variable, such as y = y(x). The derivative of such functions is at the heart of calculus and is very important and widespread in modern mathematics and science.

(#) Definition of the Derivative: If we have a function y = y(x) and its corresponding curve in the plane, consider two points on the curve. Δy is the change in y between the two points, Δx is the change in x between the two points. The ratio $\left(\frac{\Delta y}{\Delta x}\right)$ is the slope of the line through these two points. Fix one of the points, (c,y(c)), and let the other approach it along the curve, in an arbitrary way, meaning from both sides. For each position of the moving point, as it approaches the fixed point, we can calculate the slope of the line through these two points $\left(\frac{\Delta y}{\Delta x}\right)$. The limiting value of these slopes (if it exists) is called the derivative y′(c). So then we have

$$y'(c) = \lim_{\Delta x \to 0} \left(\frac{\Delta y}{\Delta x}\right)$$

as the definition of the derivative at c. y(x) is said to be differentiable on some open set (a,b) if this limit exists for all x = c in (a,b). In this definition x is the independent variable and y is the dependent variable. We say that we are taking the derivative of y with respect to x, and write this as $\frac{d}{dx}y(x)$ or $\frac{dy}{dx}$ or y′(x). We can continue to differentiate successively. For example, the second derivative y″(x) is the derivative of the first derivative y′(x), and so on with higher order derivatives. Another notation for the second derivative is $\frac{d}{dx}\left(\frac{dy}{dx}\right) = \frac{d^2y}{dx^2}$, another notation for the third derivative is $\frac{d}{dx}\left(\frac{d^2y}{dx^2}\right) = \frac{d^3y}{dx^3}$, and so on.

(#) The Meaning of the Derivative: The derivative tells about the rate of change of the function y(x) at various points. If the fixed point is (c,y(c)), then the value of the derivative at this point is y′(c). This is the slope of the unique tangent line to the curve at (c,y(c)), and is interpreted as the instantaneous rate of change of y with respect to x for the function y(x) at the point (c,y(c)). If y′(c) is negative, then the function is decreasing in an open set around x = c. If y′(c) is positive, then the function is increasing in an open set around x = c. If y′(c) = 0, then the function is neither increasing or decreasing when x = c. This generally corresponds to a local minimum or a local maximum for the function values, because at a point where y′(c) = 0, this means that the tangent line at that point is horizontal. We will spend considerable time analyzing functions using derivatives in a later section when we consider geometric applications and optimization.

(#) Calculating the Derivative: Given a certain explicit function y(x), we cannot easily calculate the derivative function from the above definition for many functions. The definition above is conceptual and meant to convey in a simple way what the derivative is. The following is the formula for calculating the derivative y′(x) from y(x):

$$y'(x) = \lim_{h \to 0} \left(\frac{y(x+h) - y(x)}{h} \right)$$

In this limit, h is taken to be a small increment of the independent variable x, called Δx, where Δx = ((x + h) - (x)) = h. The quantity $\Delta y = (y(x + h) - y(x))$ is a small increment of the dependent variable y. As h approaches 0, (x + h) approaches x, y(x + h) approaches y(x), and the slope of the secant line through the two points on the curve corresponding to x and (x + h), $\left(\frac{y(x+h) - y(x)}{h}\right)$, converges to the slope of the unique tangent line, whose value is y′(x). When considering the limit, it is often the case that the quantity $\left(\frac{y(x+h) - y(x)}{h}\right)$, called the difference quotient, approaches the common indeterminate form $\left(\frac{0}{0}\right)$. However, when y′(x) exists, this ratio does approach some limiting value, a single real number. As we learned in a previous section , we generally have to do some algebraic manipulation of the difference quotient until it is possible to figure out what the limit is. It should be noted that in the study of calculus, in its various parts, there arise many different types of indeterminate forms and we must go through some procedure to figure out what the limit in question actually is.

There is another formula that can be used to get an explicit form for the derivative of y(x), especially when we are interested in the derivative only at the one point (c,y(c)). However, the first method outlined above is generally better because it allows us to find the derivative for a larger collection of functions. Nonetheless, this formula is theoretically equivalent to the one above, and the student may actually see more clearly what the derivative means from this formulation. We will have use for it occasionally:

$$y'(c) = \lim_{x \to c} \left(\frac{y(x) - y(c)}{x - c} \right).$$

Example 1: Use both calculational formulas given above to find the derivative of y = $10x^2 + 3x - 7$ when x = 2, and compare results.

(A) With the first formula, $y'(x) = \lim_{h \to 0} \frac{y(x+h) - y(x)}{h}$

$$= \lim_{h \to 0} \frac{10(x+h)^2 + 3(x+h) - 7 - (10x^2 + 3x - 7)}{h}$$

$$= \lim_{h \to 0} \frac{10x^2 + 20xh + 10h^2 + 3x + 3h - 7 - 10x^2 - 3x + 7}{h}$$

$$= \lim_{h \to 0} \frac{20xh + 10h^2 + 3h}{h} = \lim_{h \to 0} (20x + 3 + 10h) = 20x + 3.$$

Note that until the second part of the previous line we have the indeterminate form $\left(\frac{0}{0}\right)$ when we substitute 0 for h, and so we need to continue with the algebraic manipulations. Not until we get to the point where the fractions have been eliminated can we determine the final form of the derivative function, which is y′(x) = 20x + 3. Now, for x = 2, y′(2) = 20(2)+3 = 43. Therefore the instantaneous rate of change for y(x), when x = 2, is 43 $\left(\frac{y\text{–}units}{x\text{–}units}\right)$. If for example this corresponded to a motion problem in physics where y is measured in meters and x in seconds, we would have the final result y′(2) = 43 meters/second.

(B) Now for the second formula,

$$y'(2) = \lim_{x\to 2}\left(\frac{y(x)-y(2)}{x-2}\right) = \lim_{x\to 2}\left(\frac{10x^2+3x-7-39}{x-2}\right)$$

$$= \lim_{x\to 2}\left(\frac{10x^2+3x-46}{x-2}\right) = \lim_{x\to 2}\frac{(10x+23)(x-2)}{(x-2)} = \lim_{x\to 2}(10x+23)$$

$$= 43\left(\frac{y\text{–}units}{x\text{–}units}\right).$$

Example 2: Find the derivative of y = $\frac{1}{3}x^2 - 9$ using the first definition. Then find the instantaneous rate of change of y(x) when x = -1 and x = 1.

$$y'(x) = \lim_{h\to 0}\frac{y(x+h)-y(x)}{h} = \lim_{h\to 0}\frac{\frac{1}{3}(x+h)^2-9-(\frac{1}{3}x^2-9)}{h}$$

$$= \lim_{h\to 0}\frac{\frac{1}{3}x^2+\frac{2}{3}xh+\frac{1}{3}h^2-9-\frac{1}{3}x^2+9}{h} = \lim_{h\to 0}\frac{\frac{2}{3}xh+\frac{1}{3}h^2}{h}$$

$$= \lim_{h\to 0}\left(\tfrac{2}{3}x+\tfrac{1}{3}h\right) = \tfrac{2}{3}x.$$

Therefore, y′(x) = $\frac{2}{3}x$.

Now, when x = -1, y′(-1) = $\frac{2}{3}(-1) = -\frac{2}{3}$. When x = 1, y′(1) = $\frac{2}{3}(1) = \frac{2}{3}$. The reader should recognize that y(x) is a parabola with vertex at (0,-9) that opens upward. So it is not surprising that the function is decreasing at x = -1 and increasing at x = 1. The rates of change are symmetrical because x = -1 is one unit to the left, and x = 1 is one unit to the right of the y-axis, which is the axis of symmetry for the parabola.

(#) Linearity Property of Differentiation: If y(x) is equal to some linear combination of two or more functions, such as y(x) = (a)f(x) + (b)g(x), where a and b are real numbers, then:

$$y'(x) = \lim_{h\to 0}\left(\frac{y(x+h)-y(x)}{h}\right)$$

$$= \lim_{h\to 0}\left(\frac{af(x+h)+bg(x+h)-af(x)-bg(x)}{h}\right)$$

$$= (a)\lim_{h\to 0}\left(\frac{f(x+h)-f(x)}{h}\right) + (b)\lim_{h\to 0}\left(\frac{g(x+h)-g(x)}{h}\right)$$

$$= (a)f'(x) + (b)g'(x) .$$

So we have established that the derivative of a linear combination of functions is the linear combination of the derivatives.

EXERCISES:

(1) Use both calculational formulas to find the derivative of y = $5x^2+2x$.

(2) Use the main calculational formula (the first given above) to find the derivative of y = $\frac{3}{x}$.

(3) Use the main calculational formula to find the derivative of y = $\sqrt{x}$.

(5) Derivatives of Algebraic Functions

The Algebraic functions are those that are polynomials, n^{th}-root functions, and the rational functions. These are some of the most fundamental functions that we encounter in applications of Calculus, and they most easily lend themselves to the task of deriving the basic rules that we use to calculate derivatives. We will learn rules for calculating derivatives that make the process much easier than using the limit definition of the previous chapter.

(#) The Chain Rule: In calculus, we usually encounter composite or nested functions like $y = (3x^2 - 7x)^3$ or $y = \ln(\sin(x^2 + 1))$. Therefore, we will discuss the definition of the derivative of a composite function before learning the basic techniques for calculating derivatives. For the first function from above, let $u = (3x^2 - 7x)$ and then $y = u^3$. Then y is the composite function $y(x) = y(u(x))$. For the second function from above, we let $u = (x^2 + 1)$, $v = \sin(u)$, and $y = \ln(v)$. Then y is the composite function $y(x) = y(v(u(x)))$.

In the case of only one nested function, we can write,

$$\frac{d}{dx}y(u(x)) = \lim_{\Delta x \to 0} \frac{\Delta y}{\Delta x} = \lim_{\Delta x \to 0} \left(\frac{\Delta y}{\Delta u} \cdot \frac{\Delta u}{\Delta x} \right) \quad \text{(note that } \Delta u \to 0 \text{ as } \Delta x \to 0\text{)}$$

$$= \lim_{\Delta u \to 0} \frac{\Delta y}{\Delta u} \cdot \lim_{\Delta x \to 0} \frac{\Delta u}{\Delta x} \quad \text{(if these limits exist)}$$

So: $\frac{dy}{dx} = \frac{dy}{du} \cdot \frac{du}{dx}$.

In the case of two nested functions, we can write,

$$\frac{d}{dx}y(v(u(x))) = \lim_{\Delta x \to 0} \frac{\Delta y}{\Delta x} = \lim_{\Delta x \to 0} \left(\frac{\Delta y}{\Delta v} \cdot \frac{\Delta v}{\Delta u} \cdot \frac{\Delta u}{\Delta x} \right) \quad (\Delta v \text{ and } \Delta u \to 0 \text{ as } \Delta x \to 0)$$

$$= \lim_{\Delta v \to 0} \frac{\Delta y}{\Delta v} \cdot \lim_{\Delta u \to 0} \frac{\Delta v}{\Delta u} \cdot \lim_{\Delta x \to 0} \frac{\Delta u}{\Delta x} \quad \text{(if these limits exist)}$$

So: $\frac{dy}{dx} = \frac{dy}{dv} \cdot \frac{dv}{du} \cdot \frac{du}{dx}$.

We can extend this idea to any number of nested functions, and we call this result the chain rule.

The bulk of the rest of this chapter is about the three main rules for differentiating functions: the power rule, the product rule, and the quotient rule. The chain rule is used in conjunction with these three rules. As you progress through calculus, with lots of practice, the use of the chain rule will become second nature. In this book, as in any calculus text, we will explain its use carefully at first, then soon you should be proficient.

(#) **<u>The Derivative of a Constant</u>**: Suppose y = c, where c is a real number. Then:

$$y'(x) = \lim_{h \to 0} \left(\frac{y(x+h) - y(x)}{h} \right) = \lim_{h \to 0} \left(\frac{c - c}{h} \right) = \lim_{h \to 0} \left(\frac{0}{h} \right) = 0.$$

Note that the last limit is true because h never actually reaches 0. It is the limit of that ratio that we are interested in, which is 0 for all the non-zero values that h takes on as it approaches zero.
So, we have our first rule: The derivative of a constant is zero.

(#) **<u>The Power Rule, Product Rule, Quotient Rule</u>**: Suppose $y = ax^n$ where a is a real number and n is a whole number $n \geq 0$. We wish to prove that $y'(x) = anx^{n-1}$. This is called the Power Rule. The case where n = 0 has actually been derived above when we calculated the derivative of a constant, because a constant c can be written $c(x)^0 = c(1) = c$.

We will use an induction argument to prove this result for n greater than or equal to 1. The reader is probably not familiar with a proof by mathematical induction, so I want to give a brief explanation.

When we prove something by mathematical induction, we are trying to prove a sequence of propositions $\{P(1), P(2), P(3), \ldots\}$. The proof is in two parts:

(i) Prove that the proposition is true for n = 1, that is, P(1).

(ii) Assume that P(n) is true for some n greater than or equal to 1, and prove that it is true for n+1, that is, that P(n+1) is true.

The logic being used here is that if both of these steps are done, the second part says the fact that it is true for n = 1 implies that it is true for n = 2. Then again, the second part says the fact that it is true for n = 2 implies that it is true for n = 3. Then again, the second part says the fact that it is true for n = 3 implies that it is true for n = 4, and so on. In this way we are able to conclude that the proposition is true for every whole number n from one to infinity.

Proof:

(A) Let's show that it's true when n = 1, that is, when y = ax.

$$y'(x) = \lim_{h \to 0} \left(\frac{y(x+h) - y(x)}{h} \right)$$

$$= \lim_{h \to 0} \left(\frac{a(x+h) - ax}{h} \right)$$

$$= (a) \lim_{h \to 0} \left(\frac{(x+h) - x}{h} \right),$$ because "a" doesn't depend on h.

$$= (a) \lim_{h \to 0} (1) = a(1) = a(1)x^{1-1}.$$

Therefore, we know it is true when n = 1.

(B) Now suppose it is true for some whole number n $\geq$ 1. That is

$$\frac{d}{dx}(ax^n) = (\lim_{h \to 0} \frac{a(x+h)^n - ax^n}{h}) = anx^{n-1} .$$

We must show it's true for (n + 1).

$$\frac{d}{dx}(ax^{n+1}) = \lim_{h \to 0} \frac{a(x+h)^{n+1} - a(x)^{n+1}}{h}$$

$$= \lim_{h \to 0} \frac{a(x+h)^{n}(x+h) - ax^{n}(x)}{h}$$

$$= \lim_{h \to 0} \frac{a(x+h)^{n}(x) + a(x+h)^{n}(h) - ax^{n}(x)}{h}$$

$$= (x)(\lim_{h \to 0} \frac{a(x+h)^{n} - ax^{n}}{h}) + \lim_{h \to 0} a(x+h)^{n}$$

= (x)(anx^{n-1}) + ax^{n} = anx^{n} + ax^{n}

(because we know the proposition is true for n)

= a(n + 1)x^{n}

= a(n + 1)$x^{(n+1)-1}$.

Therefore, it is true for n+1.

Therefore, by induction the proposition is true for all whole numbers $n \geq 1$. QED

This is really the first major formula for calculating derivatives. In the next few pages, we will incrementally prove this formula for functions of the form y(x) = (a)(x)^{n} , where we expand the set that n belongs to. The case of the whole numbers has already been done. Now we need to show it is true when n is a non-negative rational number, and then finally for the case where n is any rational number, positive or negative.

Summarizing, we'll state:

(**<u>The Power Rule when n is a whole number</u>**)

If $y(x) = (a)(x)^n$, where a is a real number and n is a whole number, $y'(x) = (a)(n)(x)^{n-1}$.
This has been proven above.

A few quick examples of the use of the power rule:

(a) If $y = 3x^2$, $\frac{dy}{dx} = 6x$. (Here a = 3, n = 2. Bring down "n" and multiply it by "a", then multiply that by x to the "n - 1" power).

(b) If $y = 10x^4$, $\frac{dy}{dx} = 40x^3$.

(c) If $y = 5x + 8$, $\frac{dy}{dx} = 5$.

(d) If $y = 105x^5$, $\frac{dy}{dx} = 525x^4$.

(e) If $y = \sqrt{7}x^3$, $\frac{dy}{dx} = 3\sqrt{7}x^2$.

(f) If $y = \sqrt{6}x + 16x^3 - x^6$, $\frac{dy}{dx} = \sqrt{6} + 48x^2 - 6x^5$.

(g) If $y = \frac{1}{4}x^8 - \frac{2}{3}x^2 + \frac{3}{4}x^4$, $\frac{dy}{dx} = 2x^7 - \frac{4}{3}x + 3x^3$.

(h) If $y = \frac{1}{\sqrt{2}}x^3 - 12x + \frac{13}{12}x^2$, $\frac{dy}{dx} = \frac{3}{\sqrt{2}}x^2 - 12 + \frac{13}{6}x$.

(**Use of the Chain Rule**)
We can now use the power rule (where n is a whole number) to learn how to use the chain rule. There are two cases to consider:

Case (1): This case is the most common type of situation in calculus, where we have y as an explicit function of x. Let's consider three examples of this case:

(a) We have a relation like $y = (4x^4 + 10x)^5$ where y is an explicit function of x. Make a change of variable u = $(4x^4 + 10x)$ and therefore y = u^5.

So $\frac{dy}{dx} = \frac{dy}{du} \cdot \frac{du}{dx}$

$\Rightarrow \frac{dy}{dx} = \frac{d}{du}(y) \cdot \frac{d}{dx}(u)$

$\Rightarrow \frac{dy}{dx} = \frac{d}{du}(u^5) \cdot \frac{d}{dx}(4x^4 + 10x)$

$\Rightarrow \frac{dy}{dx} = (5u^4) \cdot (16x^3 + 10)$

$= 5(4x^4 + 10x)^4 \cdot (16x^3 + 10)$

(b) We have a relation like y = $(3x^2 - 7x)^3$ where y is an explicit function of x. Make a change of variable u = $(3x^2 - 7x)$ and therefore y = u^3.

So $\frac{dy}{dx} = \frac{dy}{du} \cdot \frac{du}{dx}$

$\Rightarrow \frac{dy}{dx} = \frac{d}{du}(y) \cdot \frac{d}{dx}(u)$

$\Rightarrow \frac{dy}{dx} = \frac{d}{du}(u^3) \cdot \frac{d}{dx}(3x^2 - 7x)$

$\Rightarrow \frac{dy}{dx} = (3u^2) \cdot (6x - 7)$

$= 3(3x^2 - 7x)^2 \cdot (6x - 7)$

Note that since we now know the power rule, it is tempting to think that the derivative of y = $(3x^2 - 7x)^3$ would be just $3 \cdot (3x^2 - 7x)^2$. But that is not correct. We must multiply by the derivative of what is inside the parentheses. Conceptually, when we have a composite function y = y(u(x)) consisting of an outer and inner function, we must think of the rate of change of (y with respect to x) as the rate of change of the outer function (y with respect to u) multiplied by the rate of change of the inner function (u with respect to x).

(c) We have a relation like y = $10(x^2 - 9)^{10}$ where y is an explicit function of x. Make a change of variable u = $(x^2 - 9)$ and therefore y = $10\,u^{10}$.

So $\frac{dy}{dx} = \frac{dy}{du} \cdot \frac{du}{dx}$

$\Rightarrow \frac{dy}{dx} = \frac{d}{du}(y) \cdot \frac{d}{dx}(u)$

$\Rightarrow \frac{dy}{dx} = \frac{d}{du}(10u^{10}) \cdot \frac{d}{dx}(x^2 - 9)$

$\Rightarrow \quad \frac{dy}{dx} = 100(u)^9 \cdot (2x)$

$\quad\quad\quad = 200x(x^2 - 9)^9$

Let me say that when we have composite functions, like the last three examples, the power rule would be more correctly stated:
$\frac{d}{dx}(a(u(x))^n = (a)(n)(u(x))^{n-1} \cdot u'(x)$.

Case (2): This case is an example of what is usually called implicit differentiation, y is an implicit function of x. An important idea to be grasped here is that in the course of the calculations, when we are assuming that y is a function of x, $\frac{d}{dx}(x) = 1$ and $\frac{d}{dx}(y) = \frac{dy}{dx}$. "y" is like the quantity "in parentheses" called "u" in the examples above for case(1). Let's consider two examples of this case.

(a) We have a relation like $2y^5 + y^2 = 3x$ where y is an implicit function of x (we are assuming that y is some function of x). We can differentiate both sides with respect to x to get,

$\frac{d}{dx}(2y^5 + y^2) = \frac{d}{dx}(3x)$

$\Rightarrow \quad \frac{d}{dx}(2y^5) + \frac{d}{dx}(y^2) = 3$

$\Rightarrow \quad \frac{d}{dy}(2y^5) \cdot \frac{d}{dx}(y) \; + \; \frac{d}{dy}(y^2) \cdot \frac{d}{dx}(y) = 3$

$\Rightarrow \quad 10y^4 \cdot \frac{dy}{dx} + 2y \cdot \frac{dy}{dx} = 3$

$\Rightarrow \quad (10y^4 + 2y)\frac{dy}{dx} = 3$

So, $\frac{dy}{dx} = \frac{3}{(10y^4 + 2y)}$.

(b) We have a relation $5x^2 + 2y = 2y^5$, where we assume that y is some function of x. We differentiate both sides with respect to x,

$\frac{d}{dx}(5x^2 + 2y) = \frac{d}{dx}(2y^5)$

$\frac{d}{dx}(5x^2) + \frac{d}{dx}(2y) = \frac{d}{dx}(2y^5)$

$\Rightarrow \quad 10x + 2 \cdot \frac{d}{dx}(y) = \frac{d}{dy}(2y^5) \cdot \frac{d}{dx}(y)$

$\Rightarrow \quad 10x + 2 \cdot \frac{dy}{dx} = 10y^4 \cdot \frac{dy}{dx}$

$\Rightarrow \quad 10x = (10y^4 - 2)\frac{dy}{dx}$

So, $\frac{dy}{dx} = \frac{10x}{(10y^4 - 2)}$.

__A few quick examples of the use of the chain rule__:
(The reader should make sure that they understand these).

(a) If $y = (2x + 1)^3$, $\frac{dy}{dx} = 3(2x+1)^2(2) = 6(2x+1)^2$

(b) If $y = (17x)^4$, $\frac{dy}{dx} = 4(17x)^3(17) = 68(17x)^3$

(c) If $y = 10(12x+4)^5$, $\frac{dy}{dx} = 10(5)(12x+4)^4(12) = 600(12x+4)^4$

(d) If $y = (x-2)^{10}$, $\frac{dy}{dx} = 10(x-2)^9$ (because $\frac{d}{dx}(x-2) = 1$).

(e) If $y = (3x^4 + 2x)^5$, $\frac{dy}{dx} = 5(3x^4 + 2x)^4(12x^3 + 2)$

(f) If $y = (\sqrt{15}x - 3)^{12} + (2x+1)^3$, $\frac{dy}{dx} = 12(\sqrt{15}x - 3)^{11}(\sqrt{15}) + 3(2x+1)^2(2)$

(g) If $2y^{15} - 3y^3 - 2x + 1 = 4$,
$\Rightarrow$ $30y^{14}(\frac{dy}{dx}) - 9y^2(\frac{dy}{dx}) - 2 = 0$
$\Rightarrow$ $\frac{dy}{dx} = \frac{2}{(30y^{14} - 9y^2)}$

(h) If $y^2 + 3x^3 + 6x - 9 = 12x + y$,
$\Rightarrow$ $2y(\frac{dy}{dx}) + 9x^2 + 6 = 12 + \frac{dy}{dx}$
$\Rightarrow$ $\frac{dy}{dx} = \frac{6 - 9x^2}{2y - 1}$

Now, we should discuss the case where the power can be a positive rational number. Let m and n be positive integers. Suppose $x^m = y^n$ (so $y = x^{\frac{m}{n}}$). Differentiating with respect to x, we have, $\frac{d}{dx}(x^m) = \frac{d}{dx}(y^n)$
$\Rightarrow$ $mx^{m-1} = ny^{n-1}\frac{dy}{dx}$

Note the use of the chain rule, y is some function of x, but x is simply the independent variable.

$\Rightarrow \quad \left(\frac{m}{n}\right)x^{m-1}y^{1-n} = \frac{dy}{dx}$

$\Rightarrow \quad \left(\frac{m}{n}\right)x^{m-1}\left(x^{\frac{m}{n}}\right)^{1-n} = \frac{dy}{dx}$

$\Rightarrow \quad \left(\frac{m}{n}\right)x^{m-1}x^{\frac{m}{n}-m} = \frac{dy}{dx}$

$\Rightarrow \quad \frac{dy}{dx} = \left(\frac{m}{n}\right)(x)^{\frac{m}{n}-1}$

From this, we can now state an extended form of the power rule, where n is a non-negative rational number. To find derivatives when n is a negative rational number we can use the quotient rule, which we will consider soon.

(**The Power Rule when n is a non-negative rational number**):
If $y(x) = (a)(x)^n$, where a is a real number and n is a non-negative rational number, $y'(x) = (a)(n)(x)^{n-1}$.

Considering functions and their derivatives in a general sense, we can derive the derivative of a product of two functions and the derivative of a quotient of two functions. Then we will be able to extend the Power rule for any rational exponent, positive or negative.

(**The Product Rule**): Let u(x) and v(x) be two differentiable functions, and suppose $y(x) = u(x) \cdot v(x)$. In general,

$\Delta y = (u + \Delta u)(v + \Delta v) - (uv)$

$= (uv) + (v)(\Delta u) + (u)(\Delta v) + (\Delta u)(\Delta v) - (uv)$

$= (u)(\Delta v) + (v)(\Delta u) + (\Delta u)(\Delta v)$.

So, $y'(x) = \lim_{\Delta x \to 0}\left(\frac{\Delta y}{\Delta x}\right) = \lim_{\Delta x \to 0}\left(\frac{(u)(\Delta v) + (v)(\Delta u) + (\Delta u)(\Delta v)}{\Delta x}\right)$

$= \lim_{\Delta x \to 0} u\left(\frac{\Delta v}{\Delta x}\right) + v\left(\frac{\Delta u}{\Delta x}\right) + \left(\frac{\Delta u}{\Delta x}\right)\Delta v$

$$= \lim_{\Delta x \to 0} (u(v') + v(u') + u'(\Delta v)).$$

As $\Delta x \to 0$, $|u'(x)| < \infty$, and $\Delta v \to 0$. Therefore the third term vanishes. Therefore, we have the product rule:

If $y(x) = u(x) \cdot v(x)$, then $y'(x) = u(x)v'(x) + v(x)u'(x)$.

(**<u>The Quotient Rule</u>**): Let u(x) and v(x) be two differentiable functions, and $y(x) = \frac{u(x)}{v(x)}$. In general,

$$\Delta y = \frac{u + \Delta u}{v + \Delta v} - \frac{u}{v} = \frac{(u + \Delta u)(v) - (u)(v + \Delta v)}{v(v + \Delta v)}$$

$$= \frac{uv + v(\Delta u) - u(\Delta v) - uv}{v^2 + v(\Delta v)}$$

$$= \frac{v(\Delta u) - u(\Delta v)}{v^2 + v(\Delta v)}.$$

So, $y'(x) = \lim_{\Delta x \to 0} \left(\frac{\Delta y}{\Delta x}\right) = \lim_{\Delta x \to 0} \left(\frac{v(\frac{\Delta u}{\Delta x}) - u(\frac{\Delta v}{\Delta x})}{v^2 + v(\Delta v)}\right)$.

As, $\Delta x \to 0$, $|v(x)| < \infty$, and $\Delta v \to 0$. Therefore, $v(\Delta v)$ vanishes. Therefore, we have the quotient rule:

If $y(x) = \frac{u(x)}{v(x)}$, then $y'(x) = \left(\frac{v(\frac{du}{dx}) - u(\frac{dv}{dx})}{v^2}\right)$.

Now, suppose we have the function $y = ax^{-\frac{m}{n}}$. Assume m and n are positive whole numbers, so the exponent is a negative rational number. We can re-write this as $y = (a)(\frac{1}{x^{\frac{m}{n}}})$ and then use the quotient rule to find the derivative because we know how to find the derivative when we have a non-negative rational exponent:

The derivative is $\frac{dy}{dx} = (a)(\frac{x^{\frac{m}{n}}(0) - (1)(\frac{m}{n})x^{\frac{m}{n}-1}}{x^{\frac{2m}{n}}})$

$$= (a)(-\tfrac{m}{n})x^{(-\frac{2m}{n}+\frac{m}{n}-1)}$$
$$= (a)(-\tfrac{m}{n})x^{(-\frac{m}{n})-1}.$$

Now, we have reached our destination.

(**<u>The Power Rule when n is any rational number</u>**):

With the most recent result, we can say that for any function $y(x) = ax^n$, where a is a real number and n is any rational number, $y'(x) = an(x)^{n-1}$.

This formula is valid even when n is a real number, but we will not worry about that in this book. Most applications of the power rule involve cases where n is a rational number.

(#) <u>Additional Examples</u>:

<u>Example 1</u>: Let y = $4(3x+5)^3 + (\sqrt{2})(\sqrt{2}x-9)^4$. Find $\frac{dy}{dx}$.

$\frac{dy}{dx} = 12(3x+5)^2(3) + (\sqrt{2})(4)(\sqrt{2}x-9)^3(\sqrt{2})$

$= 36(3x+5)^2 + 8(\sqrt{2}x-9)^3$

<u>Example 2</u>: If y = $((\frac{\frac{2}{3}x^2 - \frac{1}{4}x^3}{x}) + 6x + 3)$, find $\frac{dy}{dx}$.

$\frac{dy}{dx} = \frac{d}{dx}(\frac{2}{3}x - \frac{1}{4}x^2 + 6x + 3)$

$= (\frac{2}{3} - \frac{1}{2}x + 6) = (\frac{20}{3} - \frac{1}{2}x)$.

<u>Example 3</u>: Let y = $\sqrt{13x} = (13x)^{\frac{1}{2}}$. Find $\frac{dy}{dx}$.

$\frac{dy}{dx} = (\frac{1}{2})(13x)^{\frac{1}{2}-1}(13) = (\frac{1}{2})(13x)^{-\frac{1}{2}}(13) = \frac{\sqrt{13}}{2\sqrt{x}}$.

<u>Example 4</u>: Let y = $\sqrt[3]{7x+2} = (7x+2)^{\frac{1}{3}}$. Find the derivative of y

with respect to x.

$\frac{dy}{dx} = \frac{1}{3}(7x+2)^{\frac{1}{3}-1}(7)$

$= \frac{7}{3}(7x+2)^{-\frac{2}{3}}$

Example 5: Let y = $23x^{\frac{7}{4}}$. Find the derivative of y with respect to x.

$\frac{dy}{dx} = 23(\frac{7}{4})x^{\frac{7}{4}-1}$

$= (\frac{161}{4})x^{\frac{3}{4}}$.

Example 6: If $2y = 4x^2 - 3y^3$. Find $\frac{dy}{dx}$, assuming y is a function of x.

$\frac{d}{dx}(2y) = \frac{d}{dx}(4x^2) - \frac{d}{dx}(3y^3)$

$2\frac{dy}{dx} = 8x - \left(\frac{d}{dy}(3y^3) \cdot \frac{d}{dx}(y)\right)$

$2\frac{dy}{dx} = 8x - (9y^2 \cdot \frac{dy}{dx})$

$\frac{dy}{dx} = \frac{8x}{2+9y^2}$.

Example 7: If $9(y+7)^4$ = 3x^2 , find $\frac{dy}{dx}$ assuming y is a function of x.

$\frac{d}{dx}(9(y+7)^4) = \frac{d}{dx}(3x^2)$

$\Rightarrow \quad 36(y+7)^3(\frac{dy}{dx}) = 6x$

$\Rightarrow \quad \frac{dy}{dx} = \frac{x}{6(y+7)^3}$.

Example 8: Assume that y is some function of x. If $9y^2 = 3x^3$, find $\frac{dy}{dx}$.

$\frac{d}{dx}(9y^2) = \frac{d}{dx}(3x^3)$

$\Rightarrow \quad 18y \cdot \frac{dy}{dx} = 9x^2$

$\Rightarrow \quad \frac{dy}{dx} = \frac{x^2}{2y}$.

Example 9: Assume y is a function of x, and $y^2 + 2y^3 = 14x$. Find $\frac{dy}{dx}$.

$\frac{d}{dx}(y^2) + \frac{d}{dx}(2y^3) = \frac{d}{dx}(14x)$

$2y\frac{dy}{dx} + 6y^2\frac{dy}{dx} = 14$

$\Rightarrow \quad \frac{dy}{dx} = \frac{14}{2y+6y^2}$.

Example 10: If y = $(3x+2)^8 \cdot (6x-4)^9$, find y′(x). Instead of spending hours expanding and multiplying terms, the product rule makes things much easier.

$$\begin{aligned} \tfrac{dy}{dx} &= (3x+2)^8 \cdot 9(6x-4)^8(6) + (6x-4)^9 \cdot 8(3x+2)^7(3) \\ &= (3x+2)^7(6x-4)^8 \cdot (54(3x+2)+24(6x-4)) \\ &= (3x+2)^7(6x-4)^8 \cdot (306x+12) \end{aligned}$$

Example 11: If y = $\frac{-5}{x^2}$, find y′(x) using the quotient rule.

$$\begin{aligned} \frac{dy}{dx} &= \frac{(x^2)(0)-(-5)(2x)}{(x^2)^2} \\ &= \frac{10x}{x^4} = \frac{10}{x^3} . \end{aligned}$$

Example 12: Assuming y is a function of x, and $x^2y^4+3x=4$, find $\frac{dy}{dx}$.
If $x^2y^4+3x=4$, $\quad \frac{d}{dx}(x^2y^4)+\frac{d}{dx}(3x) = \frac{d}{dx}(4)$.

To differentiate the first term we need the product rule. So,

$$((x^2)(4y^3)\tfrac{dy}{dx} + (y^4)(2x)) + 3 = 0$$
$$4x^2y^3\tfrac{dy}{dx} + 2xy^4 = (-3)$$
$$\Rightarrow \frac{dy}{dx} = \frac{-(3+2xy^4)}{4x^2y^3} .$$

Example 13: Find the tangent line to the function y = $\sqrt{x}$ when x = 4.
$\frac{dy}{dx} = \frac{1}{2\sqrt{x}}$. So, y(4) = 2 and y′(4) = $\frac{1}{2\sqrt{4}} = \frac{1}{4}$. Then $(y-2) = \frac{1}{4}(x-4)$, or $y = \frac{1}{4}x + 1$ is the tangent line to y = $\sqrt{x}$ when x = 4. This suggests that we can use this line to approximate $y = \sqrt{x}$ when x is close to 4. For example the $\sqrt{4.01} \approx \frac{1}{4}(4.01)+1 = 2.0025$. From my scientific calculator, I get 2.002498439, . This technique of using the y-values of a tangent line to

approximate the y-values of the function, at a certain point on the curve, is known as linearization, or making a linear approximation.

Example 14: Find the tangent line to y = $\frac{1}{x^2}$ when x = -2.
$\frac{dy}{dx} = \frac{-2}{x^3}$. So y(-2) = $\frac{1}{4}$ and y′(-2) = $\frac{1}{4}$. So the instantaneous rate of change of y with respect to x, when x = -2, is $\frac{1}{4}$, and $(y-\frac{1}{4}) = \frac{1}{4}(x+2)$, or $y = \frac{1}{4}x + \frac{3}{4}$, is the tangent line to y = $\frac{1}{x^2}$ when x = -2. We could use this line to approximate the values of y = $\frac{1}{x^2}$ when x is close to -2.

(#) **Derivatives of Inverse functions**: This section will deal with a topic that will help us greatly in finding derivatives of many functions.

A function f(x) from a set A to a set B is called one-to-one if for two elements x and y in A, x ≠ y implies that f(x) ≠ f(y). In addition, a function f(x) from A to B is called onto if for every y ∈ B, there is an x ∈ A, such that f(x) = y. Two functions y = f(x) and y = g(x) are inverse functions if f(x) is a one-to-one and onto function from a set A to a set B, and g(x) is a one-to-one and onto function from set B to set A, and f(g(x)) = x for all x ∈ B, and g(f(x)) = x for all x ∈ A. Sometimes we use the notation y = $f^{-1}(x)$ to denote an inverse function, in which case $f(f^{-1}(x))$ = x for all x in the domain of $f^{-1}(x)$ and $f^{-1}(f(x))$ = x for all x in the domain of f. Note that the range of f is the domain of f^{-1} and the range of f^{-1} is the domain of f. When we have a relation between x and y that is invertible, then we mean that y(x) and x(y) are inverse functions and we can then speak of $\frac{dy}{dx}$ and $\frac{dx}{dy}$.

Now, suppose y = f(x) iff x = g(y). This means f and g are inverses. Then, using the chain rule and differentiating y = f(x) with respect to y yields

$1 = f'(x) \cdot \frac{dx}{dy} \quad \Rightarrow \quad \frac{dx}{dy} = \frac{1}{f'(x)} = \frac{1}{\left(\frac{dy}{dx}\right)}$, and

differentiating x = g(y) with respect to x yields

$1 = g'(y) \cdot \frac{dy}{dx} \quad \Rightarrow \quad \frac{dy}{dx} = \frac{1}{g'(y)} = \frac{1}{\left(\frac{dx}{dy}\right)}$.

<u>Example 1</u>: Suppose we have the relation 2x + 5y = c, where c $\in$ R, and we assume that y is a function of x. Differentiating with respect to x, we have,

$\frac{d}{dx}(2x + 5y) = \frac{d}{dx}(c) \quad \Rightarrow \quad \frac{d}{dx}(2x) + \frac{d}{dx}(5y) = 0,$

$\Rightarrow \quad 2 + 5\left(\frac{dy}{dx}\right) = 0$

$\Rightarrow \quad \frac{dy}{dx} = -\frac{2}{5}$.

Now assume x is a function of y. Differentiating with respect to y, we have,

$\frac{d}{dy}(2x + 5y) = \frac{d}{dy}(c) \quad \Rightarrow \quad \frac{d}{dy}(2x) + \frac{d}{dy}(5y) = 0$

$\Rightarrow \quad 2\left(\frac{dx}{dy}\right) + 5 = 0 \quad \Rightarrow \quad \frac{dx}{dy} = -\frac{5}{2}$.

These are the derivatives $\frac{dy}{dx}$ and $\frac{dx}{dy}$, on regions of the x-y plane where x(y) and y(x) are inverse functions of each other. We can see that

$\frac{dy}{dx} = \frac{1}{\left(\frac{dx}{dy}\right)}$.

<u>Example 2</u>: Suppose we have the relation $x^3 + y^2 = 2x$, and we assume that y is some function of x. Differentiating with respect to x we have,

$\frac{d}{dx}(x^3) + \frac{d}{dx}(y^2) = \frac{d}{dx}(2x)$

$\Rightarrow \quad 3x^2 + 2y\frac{dy}{dx} = 2 \quad \Rightarrow \quad 2y\frac{dy}{dx} = (2 - 3x^2)$

$\Rightarrow \quad \frac{dy}{dx} = \left(\frac{2 - 3x^2}{2y}\right)$.

Now assume x is some function of y, differentiating with respect to y, we have,

$\frac{d}{dy}(x^3) + \frac{d}{dy}(y^2) = \frac{d}{dy}(2x)$

$\Rightarrow \quad 3x^2\frac{dx}{dy} + 2y = 2\frac{dx}{dy}$

$\Rightarrow \quad 2y = (2 - 3x^2)\frac{dx}{dy}$

$\Rightarrow \quad \frac{dx}{dy} = \left(\frac{2y}{2 - 3x^2}\right).$

These are the derivatives $\frac{dy}{dx}$ and $\frac{dx}{dy}$, on regions where x(y) and y(x) are inverse functions of each other. We can see that $\frac{dy}{dx} = \frac{1}{\left(\frac{dx}{dy}\right)}$.

(#) Summary of Rules

-If c is a constant, The $\frac{d}{dx}(c) = 0$.

-If a,b are real numbers, u(x), v(x) are differentiable functions,

$\frac{d}{dx}$ [a(u(x)) + b(v(x))] = a$\frac{d}{dx}u(x)$ + b$\frac{d}{dx}v(x)$.

-If y = y(u(x)) is a composite function,

$\frac{d}{dx}y(u(x)) = y'(u(x)) \cdot u'(x)$ or $\left(\frac{dy}{dx} = \frac{dy}{du} \cdot \frac{du}{dx}\right)$ (The Chain Rule)

-If a is a real number, and n is a rational number,

$\frac{d}{dx}(ax^n) = anx^{n-1}$. (The Power Rule)

Incorporating the chain rule into the formula, we would have:

$\frac{d}{dx}$(a)$(u(x))^n$ = an$(u(x))^{n-1} \cdot u'(x)$ (The Power Rule)

This is the most general form. $u'(x) = 1$ when $u(x) = x$, but can also happen when $u(x) = x + 3$ or something similar.

-If u(x) and v(x) are differentiable functions,

$\frac{d}{dx}(u(x) \cdot v(x)) = u(x)v'(x) + v(x)u'(x)$ (The Product Rule)

-If u(x) and v(x) are differentiable functions,

$$\frac{d}{dx}\left(\frac{u(x)}{v(x)}\right) = \frac{v(x)u'(x) - u(x)v'(x)}{(v(x))^2} \qquad \text{(The Quotient Rule)}$$

EXERCISES:

Calculate $\frac{dy}{dx}$ for each of the relations in problems 1 - 16.

(1.) y = $3x^3 - 6x^2 + 2$

(2.) y = $\sqrt{x^2 - 4}$

(3.) y = $\frac{x^2 - 9x}{\sqrt{x}}$

(4.) $x^2y^2 = x - 6$

(5.) y = $\frac{15}{x^3}$

(6.) $y^2 = x^2 - 3x + 4$

(7.) y = $\sqrt{x} \cdot (x - 8)^4$

(8.) y = $\frac{x^{\frac{3}{8}}}{x^{\frac{1}{4}}}$

(9.) y = $\frac{(x + 1)}{\sqrt[3]{x}}$

(10.) $2x + y^2 - y = 0$

(11.) $x^2 + y^2 = 9$

(12.) y = (x + 3)(x - 2)

(13.) y = $(x^2 - 16)^4$

(14.) y = $\sqrt[5]{x^6 + x^4}$

(15.) y = $2(x^5 + 4)^9$

(16.) y = x - y

(17.) Calculate $\frac{dy}{dx}$ and $\frac{dx}{dy}$ for the relation y = $x^2 + 3$ and verify that they are reciprocals.

(18.) Find the equation of the line tangent to $y = \sqrt{4x}$ at x = 5. Interpret the slope of this line.

(19.) Find the equation of the line tangent to $y = 3 - x^2$ at x = $\frac{3}{2}$. Interpret the slope of this line.

(20.) Calculate y′(1), y′(5), y′(10) for y = $\frac{3}{x}$. What does this say about the rate of change of this function as x increases to (+∞)?

(21.) Calculate y′$\left(\frac{1}{2}\right)$, y′$\left(\frac{1}{5}\right)$, and y′$\left(\frac{1}{10}\right)$ for y = $\frac{3}{x}$. What does this say about the rate of change of y(x) as x approaches 0 from the right?

(6) Derivatives of Transcendental Functions

The Transcendental functions are the logarithmic and exponential, trigonometric and inverse trigonometric functions. The algebraic and transcendental functions included in this book, in fact do cover nearly all of the common functions encountered in calculus. There are other functions considered to be transcendental that are often treated in more extensive books on calculus, namely the hyperbolic and inverse hyperbolic functions. We will not consider these other functions, since this book is of a more introductory nature. The omission of a treatment of these functions doesn't diminish the nature of what we are presenting here.

(#) Rolle's Theorem and the Mean Value Theorem: Our first task is to learn the Mean Value theorem (MVT), a very important theorem in calculus. But to do so we need to know a theorem that is commonly used to help us prove it, namely Rolle's theorem. Let's first state what Rolle's theorem is.

(**Rolle's Theorem**): If y(x) is continuous on a closed interval [a,b], differentiable on the open interval (a,b), and y(a) = y(b), then there exists a number c $\in$ (a,b) such that y′(c) = 0.

Proof:
Let y(x) be a function that satisfies the hypotheses of the theorem.
We can ignore the trivial case of constant functions on [a,b]. Let's suppose that there does not exist such a number c. If there is some number d $\in$ (a,b) such that y(d) $\neq$ y(a), then since there is no number c in (a,b) such

that $y'(c) = 0$, we can only conclude that y(x) is monotone increasing or decreasing on [a,b]. If this is the case, then certainly we can conclude that $y(b) \neq y(a)$. But this is a contradiction of our hypothesis. Therefore, there must be some c in (a,b) such that $y'(c) = 0$. QED

The Mean Value theorem (MVT) is a more general version of Rolle's theorem. We do not require that f(a) = f(b).

(**The Mean Value Theorem**): If y(x) is continuous on [a,b] and differentiable on (a,b), then there exists a number $c \in (a,b)$ such that $y'(c) = \left(\frac{y(b) - y(a)}{b - a}\right)$.

Proof:

Define $d(x) = y(x) - [\, y(a) + \left(\frac{y(b) - y(a)}{b - a}\right)(x - a)\,]$. Then d(x) is a function that satisfies the hypotheses of the theorem, and d(a) = 0 and d(b) = 0. Therefore, Rolle's theorem says that there exists a number c in (a,b) such that $d'(c) = y'(c) - \left(\frac{y(b) - y(a)}{b - a}\right) = 0$. That is $y'(c) = \left(\frac{y(b) - y(a)}{b - a}\right)$. QED

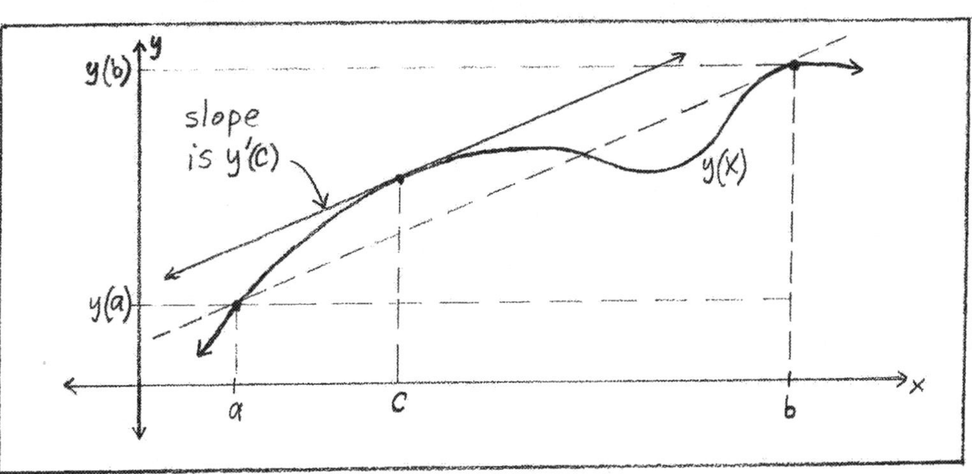

There is an important point to be made about the relationship between differentiability (where the derivative exists) and continuity (where the function is continuous). Let y(x) be a differentiable function on (a,b) and continuous on [a,b]. Let c be any number in (a,b), and let $x \in (a,b)$ and

close to c. The mean value theorem says that there exists a number d between x and c such that $\left(\frac{y(x)-y(c)}{x-c}\right) = y'(d)$. Since y(x) is differentiable on (a,b), then we know that y'(d) is a finite number. Now, we have (y(x) - y(c)) = y'(d)$\cdot$(x - c), so that when we take the limit of both sides as x$\rightarrow$c, we get $\lim_{x \to c}$ (y(x) - y(c)) = $\lim_{x \to c} (y'(d)) \cdot (x-c) = 0$. Therefore we have that the $\lim_{x \to c} y(x) = y(c)$, which says that y(x) is continuous at x = c. So if a function is differentiable at a point, then it is also continuous there. However, the converse is not true. As an example consider the function y = |x|, which is continuous at x = 0, but not differentiable at x = 0.

(#) L'Hospital's Rule: This is a result that makes the evaluation of limits much easier in a lot of situations. This theorem is true only for the indeterminate forms $\left(\frac{0}{0}\right)$ and $\left(\frac{\infty}{\infty}\right)$. We will prove it only for the indeterminate form $\left(\frac{0}{0}\right)$. The statement of the theorem is:
Let f(x) and g(x) be differentiable functions on the set (a,c)$\cup$(c,b), where a < c < b, and the $\lim_{x \to c} \frac{f(x)}{g(x)} = \frac{0}{0}$. Then $\lim_{x \to c} \frac{f(x)}{g(x)} = \lim_{x \to c} \frac{f'(x)}{g'(x)}$.

Proof:
Let x approach c, and let Δx be the distance between x and c. Let $\frac{\Delta f}{\Delta x}$ be the slope of the line between the two points $(x, f(x))$ and $(c, 0)$, and let $\frac{\Delta g}{\Delta x}$ be the slope of the line between the two points $(x, g(x))$ and $(c, 0)$. From the MVT, there exist x-values d and e between x and c, such that $\frac{\Delta f}{\Delta x}$ = f'(d) and $\frac{\Delta g}{\Delta x}$ = g'(e). Therefore,

$$\lim_{x \to c} \frac{f(x)}{g(x)} = \lim_{x \to c} \frac{\left(\frac{\Delta f}{\Delta x}\right)}{\left(\frac{\Delta g}{\Delta x}\right)} = \lim_{x \to c} \frac{f'(d)}{g'(e)} = \lim_{x \to c} \frac{f'(x)}{g'(x)}.$$ QED

(#) A Limit Theorem for the Natural Exponential Function:
A useful limit in mathematics is the following: $\lim_{n \to \infty} \left(1+\frac{x}{n}\right)^n = e^x$.
This result will make it much easier to find the derivative of the

natural logarithmic function ln(x) in the next section.

Proof:

Note that we have the indeterminate form 1^{∞}. We will change the indeterminate form to $(0 \cdot \infty)$ and then to $\left(\frac{0}{0}\right)$ so we can use L'Hospital's Rule to evaluate it. Let y = $\left(1 + \frac{x}{n}\right)^{n}$.

We will find the $\lim_{n\to\infty} ln(y) = \lim_{n\to\infty} ln\left(1 + \frac{x}{n}\right)^{n} = \lim_{n\to\infty}(n)(ln\left(1 + \frac{x}{n}\right))$, using a rule of logarithms. This is a $(0 \cdot \infty)$ indeterminate form.

Rewrite this expression as the $\lim_{n\to\infty}(n)(ln\left(1 + \frac{x}{n}\right)) = \lim_{n\to\infty} \frac{ln(1 + \frac{x}{n})}{(\frac{1}{n})}$.

Now we have a $\left(\frac{0}{0}\right)$ indeterminate form and we can use L'Hospital's Rule to evaluate it. Differentiating with respect to n,

$$\lim_{n\to\infty} \frac{ln(1 + \frac{x}{n})}{(\frac{1}{n})} = \lim_{n\to\infty} \frac{\frac{1}{(1+\frac{x}{n})} \cdot \left(\frac{-x}{n^2}\right)}{\left(\frac{-1}{n^2}\right)} = \lim_{n\to\infty} \frac{x}{(1 + \frac{x}{n})} = \frac{x}{1} = x.$$

Recall, that this is the $\lim_{n\to\infty} ln(y)$. So $\lim_{n\to\infty}(y) = e^{(\lim_{n\to\infty} ln(y))} = e^{x}$.

QED

(#) <u>Derivative of Natural Logarithm and Exponential Functions</u>:

We are now in a position to derive the derivatives of y = ln(x) and y = e^{x}.

Using the definition of the derivative (we can assume h > 0),

$$\frac{d}{dx}(ln(x)) = \lim_{h\to 0}\left(\frac{ln(x+h) - ln(x)}{h}\right) = \lim_{h\to 0}\left(\frac{1}{h}\right)\left(ln\left(\frac{x+h}{x}\right)\right)$$

$$= \lim_{h\to 0}\left(\frac{1}{h}\right)\left(ln\left(1 + \frac{h}{x}\right)\right) = \lim_{h\to 0} ln\left(1 + \frac{h}{x}\right)^{\frac{1}{h}}$$

Now let, n = $\frac{1}{h}$, h = $\frac{1}{n}$.

Then $\frac{d}{dx}(ln(x)) = \lim_{n\to\infty} ln\left(1 + \frac{(\frac{1}{x})}{n}\right)^{n}$.

Using our limit theorem for e^x,

$$\frac{d}{dx}(ln(x)) = ln\left(e^{\left(\frac{1}{x}\right)}\right) = \frac{1}{x}.$$

Now use the fact that the natural exponential function and the natural logarithmic function are inverse functions, which we can express as, y = e^x iff x = ln(y). Let's differentiate x = ln(y) with respect to x to get 1 = $\left(\frac{1}{y}\right)\frac{dy}{dx}$. So $\frac{dy}{dx} = y$, or $\frac{d}{dx}(e^x) = e^x$.

Therefore: $\frac{d}{dx}(ln(x))$ = $\frac{1}{x}$ and $\frac{d}{dx}(e^x)$ = e^x

(#) <u>Derivative of General Logarithmic and Exponential Functions</u>:

There are formulas for the derivatives of y = $log_a(x)$ and y = a^x which can easily be derived from the derivatives of the natural logarithmic and exponential functions. Note that allowable values for a are a > 0, a ≠ 1.

(A) y = $a^x = e^{ln(a^x)} = e^{x(lna)} = e^{(lna)x}$.

So $\frac{dy}{dx}$ = $e^{(lna)x}(lna) = a^x(lna)$.

(B) y = $log_a(x)$ = $\frac{ln(x)}{lna}$, using a rule of logarithms.

So $\frac{dy}{dx} = \left(\frac{1}{lna}\right) \cdot \left(\frac{1}{x}\right)$.

(#) <u>Trigonometric Functions and Their Inverses</u>: We will now turn our attention to the trigonometric and inverse trigonometric functions. Once we have the derivatives of these functions, we'll have the derivatives of the majority of the functions which are important in calculus and we can turn our attention to applications of the derivative.

There is a limit formula in calculus which is very common and useful. It is usually just stated in most calculus books, with a rather complicated geometric proof that we will not worry about here, at the beginning of the consideration of limits. (Since the student now knows L'Hospital's rule, they can verify it for themselves soon, once you know the derivative of sin(x)). It is,

$$\lim_{x \to 0} \left(\frac{sinx}{x}\right) = 1 \ .$$

One by one, we will go through each of the six trigonometric functions and the six corresponding inverse trigonometric functions. I will use "wrt" to mean "with respect to." It will help the reader to recall the trigonometric identities:

$$sin(u \pm v) = sin(u)cos(v) \pm sin(v)cos(u)$$
$$cos(u \pm v) = cos(u)cos(v) \mp sin(u)sin(v)$$
$$sin^2(u) + cos^2(u) = 1$$
$$1 + tan^2(u) = sec^2(u)$$
$$1 + cot^2(u) = csc^2(u)$$

(**<u>The Trigonometric Functions</u>**):

(A) y = sin(x): $\frac{dy}{dx} = \lim_{h \to 0} \frac{sin(x+h) - sin(x)}{h}$

$$= \lim_{h \to 0} \frac{sin(x)cos(h) + cos(x)sin(h) - sin(x)}{h}$$

$$= \cos(x) \cdot \lim_{h \to 0} \frac{sin(h)}{h} = \cos(x).$$

(B) y = cos(x): $\frac{dy}{dx} = \lim_{h \to 0} \frac{cos(x+h) - cos(x)}{h}$

$$= \lim_{h \to 0} \frac{cos(x)cos(h) - sin(x)sin(h) - cos(x)}{h}$$

$$= \text{-sin(x)} \cdot \lim_{h \to 0} \frac{sin(h)}{h} = (\text{-sin(x)} \cdot 1) = -sin(x).$$

(C) y = tan(x): Using the quotient rule,

$$\frac{dy}{dx} = \frac{d}{dx}\frac{sin(x)}{cos(x)} = \frac{cos(x)cos(x) - sin(x)(-sin(x))}{cos^2(x)}$$

$$= \frac{(cos^2(x) + sin^2(x))}{cos^2(x)}$$

$$= \frac{1}{cos^2(x)} = sec^2(x).$$

(D) y = cot(x): Using the quotient rule,

$$\frac{dy}{dx} = \frac{d}{dx}\frac{cos(x)}{sin(x)} = \frac{sin(x)(-sin(x)) - cos(x)cos(x)}{sin^2(x)}$$

$$= \frac{-(sin^2(x) + cos^2(x))}{sin^2(x)}$$

$$= \frac{-1}{sin^2(x)} = -csc^2(x).$$

(E) y = sec(x): Using the quotient rule,

$$\frac{dy}{dx} = \frac{d}{dx}\frac{1}{cos(x)} = \frac{cos(x)(0) - (1)(-sin(x))}{cos^2(x)}$$

$$= \frac{sin(x)}{cos^2(x)}$$

$$= \left(\frac{1}{cos(x)}\right)\left(\frac{sin(x)}{cos(x)}\right)$$

$$= \text{sec(x)tan(x)}.$$

(F) y = csc(x): Using the quotient rule,

$$\frac{dy}{dx} = \frac{d}{dx}\frac{1}{sin(x)} = \frac{sin(x)(0) - (1)cos(x)}{sin^2(x)}$$

$$= \frac{-cos(x)}{sin^2(x)}$$

$= \left(\frac{-1}{sin(x)}\right)\left(\frac{cos(x)}{sin(x)}\right)$

$= -csc(x)cot(x)$.

(**The Inverse Trigonometric Functions**):

(A) $y = sin^{-1}(x)$ iff $x = sin(y)$

Differentiating $x = sin(y)$ wrt x, we get,

$1 = cos(y) \cdot \frac{dy}{dx} \Rightarrow \frac{dy}{dx} = \frac{1}{cos(y)}$

$= \frac{1}{\sqrt{(1 - sin^2(y)}}$

$= \frac{1}{\sqrt{1 - x^2}}$.

(B) $y = cos^{-1}(x)$ iff $x = cos(y)$

Differentiating $x = cos(y)$ wrt x, we get,

$1 = -sin(y) \cdot \frac{dy}{dx} \Rightarrow \frac{dy}{dx} = \frac{-1}{sin(y)}$

$= \frac{-1}{\sqrt{(1 - cos^2(y)}}$

$= \frac{-1}{\sqrt{1 - x^2}}$.

(C) $y = tan^{-1}(x)$ iff $x = tan(y)$

Differentiating $x = tan(y)$ wrt x, we get,

$1 = sec^2(y) \cdot \frac{dy}{dx} \Rightarrow \frac{dy}{dx} = \frac{1}{sec^2(y)}$

$= \frac{1}{1 + tan^2(y)}$

$= \frac{1}{1 + x^2}$.

(D) $y = cot^{-1}(x)$ iff $x = cot(y)$

Differentiating $x = cot(y)$ wrt x, we get,

$$1 = -csc^2(y) \cdot \frac{dy}{dx} \Rightarrow \frac{dy}{dx} = \frac{-1}{csc^2(y)}$$

$$= \frac{-1}{1 + cot^2(y)}$$

$$= \frac{-1}{1 + x^2}.$$

(E) $y = sec^{-1}(x)$ iff $x = sec(y)$

Differentiating x = sec(y) wrt x, we get,

$$1 = sec(y)tan(y) \cdot \frac{dy}{dx} \Rightarrow \frac{dy}{dx} = \frac{1}{sec(y)tan(y)}$$

$$= \frac{1}{sec(y)\sqrt{(sec^2(y) - 1}}$$

$$= \frac{1}{x\sqrt{x^2 - 1}}.$$

(F) $y = csc^{-1}(x)$ iff $x = csc(y)$

Differentiating x = csc(y) wrt x, we get,

$$1 = -csc(y)cot(y) \cdot \frac{dy}{dx} \Rightarrow \frac{dy}{dx} = \frac{-1}{csc(y)cot(y)}$$

$$= \frac{-1}{csc(y)\sqrt{csc^2(y) - 1}}$$

$$= \frac{-1}{x\sqrt{x^2 - 1}}.$$

The chain rule is of course very important with all of the functions that we have been studying in this chapter as well. These can be established in each case just as we did in the previous chapter with the examples there which involved algebraic functions. Let me illustrate with two examples:

(a) Let $y = sin(12x^2 + 2x)$. Then if we let $u = (12x^2 + 2x)$, then y = sin(u).

So, $\frac{dy}{dx} = \frac{dy}{du} \cdot \frac{du}{dx}$,

$$\frac{dy}{dx} = \frac{d}{du}(sin(u)) \cdot \frac{d}{dx}(12x^2 + 2x)$$

$$\frac{dy}{dx} = cos(u) \cdot (24x + 2)$$

$$\frac{dy}{dx} = cos(12x^2 + 2x) \cdot (24x + 2)$$

We would then have the formula: $\frac{d}{dx}(sin\ u(x)) = cos(u(x)) \cdot u'(x)$.

This would be a most general formula.

(b) Let $y = 10e^{(3x^4+1)}$. Then if we let $u = (3x^4 + 1)$, then $y = 10e^u$.

So, $\frac{dy}{dx} = \frac{dy}{du} \cdot \frac{du}{dx}$,

$\frac{dy}{dx} = \frac{d}{du}(10e^u) \cdot \frac{d}{dx}(3x^4 + 1)$

$\frac{dy}{dx} = 10e^u \cdot (12x^3)$

$\frac{dy}{dx} = 10e^{(3x^4+1)} \cdot (12x^3)$

$= 120x^3 \cdot e^{(3x^4+1)}$

We would then have the formula: $\frac{d}{dx}(e^{(u(x))}) = e^{(u(x))} \cdot u'(x)$.

This would be a most general formula.

(**<u>Some quick examples of the use of the chain rule</u>**):

The reader should strive to understand all of these examples, because unfortunately calculus can be a difficult subject.

(a) If $y = sin(3x)$, $\frac{dy}{dx} = (3)cos(3x)$

(b) If $y = cos(4x^2)$, $\frac{dy}{dx} = -sin(4x^2) \cdot (8x) = -8x\ sin(4x^2)$

(c) If $y = tan(3x + 1)$, $\frac{dy}{dx} = (3)sec^2(3x + 1)$

(d) If $y = sin(8x^3) + cos(8x^3)$, $\frac{dy}{dx} = (24x^2)\ cos(8x^3) - (24x^2)\ sin(8x^3)$

$= (24x^2)(cos(8x^3) - sin(8x^3))$

(e) If $y = csc(-2\pi x)$, $\frac{dy}{dx} = -csc(-2\pi x)cot(-2\pi x) \cdot (-2\pi)$

$= (2\pi)\ csc(-2\pi x)cot(-2\pi x)$

(f) If $y = sin^{-1}(14x)$, $\frac{dy}{dx} = \frac{1}{\sqrt{1-(14x)^2}} \cdot (14) = \frac{14}{\sqrt{1-(14x)^2}}$

(g) If $y = tan^{-1}(ax^2)$, $\frac{dy}{dx} = \frac{1}{1+(ax^2)^2} \cdot (2ax) = \frac{2ax}{1+a^2x^4}$

(h) If $y = sin(13x) - cos(15x) + tan(3x^2 + x)$,

$\frac{dy}{dx} = 13\ cos(13x) + 15\ sin(15x) + (sec^2(3x^2 + x)) \cdot (6x + 1)$.

(i) If $y = e^{mx} + ln(9x^2 + 6x + 1)$, $\frac{dy}{dx} = me^{mx} + \frac{1}{(9x^2 + 6x + 1)} \cdot (18x + 6)$

(j) If $y = e^{5x}\ sin(2x)$, $\frac{dy}{dx} = (e^{5x})(2cos(2x)) + sin(2x)(5e^{5x})$

(k) If $y = \frac{ln(6x)}{tan(8x+2)}$, $\frac{dy}{dx} = \frac{tan(8x+2)(\frac{6}{6x}) - ln(6x)(sec^2(8x+2))(8)}{tan^2(8x+2)}$

$= \frac{(\frac{tan(8x+2)}{x}) - (8)ln(6x)(sec^2(8x+2))}{tan^2(8x+2)}$

(l) If $y = cot(20x)$, $\frac{dy}{dx} = -csc^2(20x) \cdot 20 = -20csc^2(20x)$

(#) Summary of Rules:

Note that we are not stating them in their most general form, because these basic forms are useful enough when the reader understands how to use the chain rule.

$\frac{d}{dx}(log_a(x)) = \frac{1}{x\ (lna)}$

$\frac{d}{dx}\ ln(x) = \frac{1}{x}$

$\frac{d}{dx}(a^x) = a^x\ (lna)$

$\frac{d}{dx}\ e^x = e^x$

$\frac{d}{dx}sin(x) = cos(x)$

$\frac{d}{dx}tan(x) = sec^2(x)$

$\frac{d}{dx}sec(x) = sec(x)tan(x)$

$\frac{d}{dx}cos(x) = -sin(x)$

$\frac{d}{dx}cot(x) = -csc^2(x)$

$\frac{d}{dx}csc(x) = -csc(x)cot(x)$

$\frac{d}{dx}sin^{-1}(x) = \frac{1}{\sqrt{1-x^2}}$

$\frac{d}{dx}tan^{-1}(x) = \frac{1}{1+x^2}$

$\frac{d}{dx}sec^{-1}(x) = \frac{1}{x\sqrt{x^2-1}}$

$\frac{d}{dx}cos^{-1}(x) = \frac{-1}{\sqrt{1-x^2}}$

$\frac{d}{dx}cot^{-1}(x) = \frac{-1}{1+x^2}$

$\frac{d}{dx}csc^{-1}(x) = \frac{-1}{x\sqrt{x^2-1}}$

(#) <u>Additional Examples</u>:

<u>Example 1</u>: Find the derivative of y = $e^{\sqrt{x}}$.

Let u = $\sqrt{x}$, then y = e^u. Then using the chain rule,

$$\frac{dy}{dx} = \frac{dy}{du} \cdot \frac{du}{dx} = \frac{d}{du}(y(u)) \cdot \frac{d}{dx}(u(x))$$
$$= \frac{d}{du}(e^u) \cdot \frac{d}{dx}(\sqrt{x})$$
$$= e^u \cdot \frac{1}{2\sqrt{x}}$$
$$= e^{\sqrt{x}}\left(\frac{1}{2\sqrt{x}}\right) .$$

<u>Example 2</u>: Find the derivative of y = ln($7x^3 + 13x^2 - 9x$).

Let u = $(7x^3 + 13x^2 - 9x)$, then y = ln(u). Then using the chain rule,

$$\frac{dy}{dx} = \frac{dy}{du} \cdot \frac{du}{dx} = \frac{d}{du}(y(u)) \cdot \frac{d}{dx}(u(x))$$
$$= \frac{d}{du}(ln(u)) \cdot \frac{d}{dx}(7x^3 + 13x^2 - 9x)$$
$$= \frac{1}{u} \cdot (21x^2 + 26x - 9)$$
$$= \left(\frac{21x^2 + 26x - 9}{7x^3 + 13x^2 - 9x}\right).$$

<u>Example 3</u>: Suppose we have the relation $e^{xy} + y^2 = 10$, find $\frac{dy}{dx}$.

Assuming that y is a function of x, and differentiating wrt x,

$$(e^{xy})\left(x\frac{dy}{dx} + y(1)\right) + 2y \cdot \frac{dy}{dx} = 0$$
$$\Rightarrow \quad xe^{xy}\frac{dy}{dx} + ye^{xy} + 2y \cdot \frac{dy}{dx} = 0$$
$$\Rightarrow \quad (2y + xe^{xy}) \cdot \frac{dy}{dx} = (-y)e^{xy} \quad \Rightarrow \quad \frac{dy}{dx} = \frac{-ye^{xy}}{(2y + xe^{xy})}$$

Example 4: Suppose we have $y - sin(xy) = 1$, find $\frac{dy}{dx}$. Once again here we are assuming that y is some function of x. Differentiating wrt x,

$$\frac{dy}{dx} - (cos(xy)(x\frac{dy}{dx} + y(1)) = 0$$
$$\Rightarrow \quad (1 - xcos(xy)) \cdot \frac{dy}{dx} = ycos(xy)$$
$$\Rightarrow \quad \frac{dy}{dx} = \frac{ycos(xy)}{(1 - xcos(xy))}$$

Example 5: Find $\frac{dy}{dx}$ if $y = tan^3(\pi x^2)$.

$$\frac{dy}{dx} = 3(tan^2(\pi x^2)) \cdot (sec^2(\pi x^2)) \cdot (2\pi x)$$
$$= 6\pi x \cdot (tan^2(\pi x^2)) \cdot (sec^2(\pi x^2))$$

Example 6: Find $\frac{dy}{dx}$ if $y = cos(19x^3 + x) + csc(\pi x)$.

$$\frac{dy}{dx} = -sin(19x^3 + x) \cdot (57x^2 + 1) + (-csc(\pi x)cot(\pi x)) \cdot (\pi)$$
$$= -(57x^2 + 1) \cdot sin(19x^3 + x) - (\pi) \cdot (csc(\pi x)cot(\pi x))$$

Example 7: Find $\frac{dy}{dx}$ if y = $2^x(x^2 + 1)$.

$$\frac{dy}{dx} = 2^x \cdot (2x) + (x^2 + 1) \cdot (2^x(ln2))$$
$$= 2x\,(2^x) + (ln2)\,(x^2 + 1)\,2^x \ .$$

Example 8: Suppose y = $\frac{log_5(x^2)}{x - x^3}$, find $\frac{dy}{dx}$.

$$\frac{dy}{dx} = \frac{(x - x^3)\left(\frac{1}{(ln5)\,x^2}\right)(2x) - (log_5(x^2))(1 - 3x^2)}{(x - x^3)^2}$$
$$= \frac{(\frac{2}{ln5})(1 - x^2) + (3x^2 - 1)((log_5(x^2))}{(x - x^3)^2}$$

Example 9: Suppose y = $sin^{-1}(2x)$, find $\frac{dy}{dx}$.

$$\frac{dy}{dx} = \frac{1}{\sqrt{1 - (2x)^2}} \cdot (2) = \frac{2}{\sqrt{1 - 4x^2}} \ .$$

Example 10: Suppose y = $tan^{-1}(9x^2 - 4)$, find $\frac{dy}{dx}$.

$$\frac{dy}{dx} = \frac{1}{1 + (9x^2 - 4)^2} \cdot (18x) = \left(\frac{18x}{1 + (9x^2 - 4)^2}\right) .$$

Example 11: Suppose y = $csc^{-1}(e^x)$, find $\frac{dy}{dx}$.

$$\frac{dy}{dx} = \left(\frac{-1}{e^x\sqrt{(e^x)^2 - 1}}\right) \cdot (e^x) = \frac{-1}{\sqrt{e^{2x} - 1}} .$$

Example 12: If we have y = $\frac{e^{(x^2 - 1)}}{ln(x)}$, find $\frac{dy}{dx}$.

$$\frac{dy}{dx} = \frac{(ln(x)) \cdot e^{(x^2 - 1)}(2x) - e^{(x^2 - 1)}(\frac{1}{x})}{(ln(x))^2}$$

$$= \left(e^{(x^2 - 1)}\right)\left(\frac{2x}{ln(x)} - \frac{1}{x(ln(x))^2}\right) .$$

Example 13: Find the tangent line to $y = e^x$ at the point (0,1).
$\frac{dy}{dx} = e^x$, so y'(0) = 1. Therefore $(y - 1) = (1)(x - 0)$. So the line y = (x + 1) is tangent to y = e^x at (0,1). This result allows us to get a good approximation to y = e^x, when x is close to 0. For example, for x = 0.03, $e^{(0.03)} \approx 1.03$ using the tangent line. On my scientific calculator, I get $e^{(0.03)} = 1.030454534$. The tangent line provides us with a good quick approximation.

Example 14: Find the tangent line to y = ln(x) at the point (1,0).
$\frac{dy}{dx} = \frac{1}{x}$. So, y'(1) = 1. Therefore $(y - 0) = (1)(x - 1)$. So the line $y = (x - 1)$ is tangent to y = ln(x) at (1,0). This result allows us to get a good approximation to y = ln(x), when x is close to 1. For example, for x = 1.025, ln(1.025) $\approx$ 0.025 is a quick approximation using the tangent line. On my scientific calculator, ln(1.025) = 0.024692612 .

EXERCISES:

Calculate $\frac{dy}{dx}$ for each of the following functions (1 - 20).

(1.) $y = e^{3x+1}$

(2.) $y = ln(9x^2)$

(3.) $y = log_4(4x^2 + 2)$

(4.) $y = 5^x$

(5.) $y = x^2e^{-x}$

(6.) $y = \frac{ln(x)}{x^3}$

(7.) $y = sin(\sqrt{x}\,)$

(8.) $y = cos(sin(2x))$

(9.) $y = tan(x - cos(x))$

(10.) $y = x \cdot csc(x)$

(11.) $y = 2^{(3x^2 + 4x + 2)}$

(12.) $y = cot(x) \cdot tan(x)$

(13.) $y = (x)(4^{-x})$

(14.) $y = \frac{1}{\sqrt{2\pi}} \cdot e^{-\frac{x^2}{2}}$

(15.) $y = sin^2(x)$

(16.) $y = sec^3(3x)$

(17.) $y = sin^{-1}(x + 4)$

(18.) $y = sec^{-1}(10x^7)$

(19.) $y = tan^{-1}(e^{2x})$

(20.) $y = cot^{-1}(2x^2)$

(21.) Find the equation of the line tangent to y = e^{3x} when x = 2. Interpret the slope of this line.

(22.) Find the equation of the line tangent to $y = ln(x - 1)$ when x = 7. Interpret the slope of this line.

(23.) Calculate y'(0), y'($\frac{\pi}{6}$), and y'($\frac{\pi}{3}$) for y = sin(x). What does this say about the rate of change of y as x approaches $\frac{\pi}{2}$ from the left?

(7) Applications of The Derivative

(#) **Curve Analysis**: In this section we will analyze the graphs of several functions, that is, the function's curve in the x-y coordinate plane. The tools that we will use are primarily the first derivative y'(x) and the second derivative y''(x), vertical and horizontal asymptotes, and the x and y intercepts. One of the important parts of curve analysis is the analysis of local minimums and maximums, which is important in optimization problems.

(**The Derivative and Critical Values**): The meaning of the value of the derivative y'(c) of a function y(x) at a certain point on the curve (c,y(c)) is interpreted as the instantaneous rate of change of the function at that point, represented by the slope of the unique tangent line at that point. y'(x) gives us information about the rate of change at the various points on the curve of y(x). y''(x) gives us information about the rate of change of y'(x) at the various points on the curve. This idea extends to higher derivatives, though it is the case that only the first and second derivatives are of primary use to us here. The derivative fails to exist at a point (c,y(c)) on the curve of y(x) if there is no unique tangent line there. This would occur where the curve of y(x) has a cusp (a sharp point), or where there is a vertical asymptote. The function would not even be defined for the x-value of a vertical asymptote. We call the curve smooth on an open interval where the derivative is defined and continuous, and we would say the function is differentiable on that open interval. For a given function y(x), we call the collection of numbers $\{c_i\}$ where y'(c) = 0, or where y'(c) does not exist the critical values of y(x). The critical values break the domain of y(x) into consecutive open intervals. On each such open interval the function is decreasing (y′(x) < 0), increasing (y′(x) > 0), or constant (y′(x) = 0) for all x-values in that interval.

(**Inflection Values**): We call the collection of points $\{d_i\}$ where y″(d) = 0 or y″(d) does not exist the inflection values of y(x). The inflection values break the domain of y(x) into a different set of consecutive open intervals. On each such open interval the function is concave down (y″(x) < 0), concave up (y″(x) > 0), or zero in concavity (y″(x) = 0) for all x-values in that interval. Concavity tells us about the way that the function y(x) is bending at all x-values in each of the open intervals defined by the inflection values $\{d_i\}$. This is of course the way that y′(x) is changing in the interval. If in an open interval, y′(x) is continuously becoming smaller (for example going from positive numbers toward negative numbers), then the concavity is negative, and a negatively concave function is bending downward as we proceed from left to right. If in an open interval, y′(x) is continuously becoming larger (for example going from negative numbers toward positive numbers), then the concavity is positive, and a positively concave function is bending upward as we proceed from left to right.

(**Minimums and Maximums**): Consider the open intervals defined from the inflection values $\{d_i\}$. For a smooth function on one of these open intervals containing the critical value c, if y′(c) = 0 at the point (c,y(c)) and the function is negatively concave for all points in the interval, then there is a local maximum at (c,y(c)). If y′(c) = 0 at the point (c,y(c)) and the function is positively concave for all points in the interval, then there is a local minimum at (c,y(c)). We say local, rather than absolute, because we can only speak about what is happening in a small region around x = c. There may be many local minimums or local maximums in the graph of a given function over its entire domain.

An absolute minimum or maximum is the absolute lowest or highest that a function reaches on a given interval, usually a closed interval, but it could be an open interval. We would consider the function values at each of the local extrema (minimums and maximums) that are contained within the interval, as well as the function values at both of the end points of an interval (if the end point is included). If we have an end point that is not included, there will be no absolute minimum or maximum at that end point, even in the case that the function values may be approaching a certain value. However, an absolute minimum or maximum often occurs at the boundaries of an interval like [a,b], and many times not at any local minimums or maximums within (a,b).

(**Linear Functions**): A linear function is one of the form y = mx + b, where m and b are constants. Its domain is all real numbers.

y'(x) = m, for all x-values. Therefore the tangent line for a linear function, at any x-value, is the linear function itself with slope m. If m is negative, the line is always decreasing as we go from left to right. If m is positive, the line is always increasing as we go from left to right.

y"(x) = 0 for all x-values. Therefore, a line never bends in an upward or downward way as we proceed from left to right. Another way of thinking of this is that the slope y'(x) never changes.

A line y(x) is defined for all x and never bends in any way, so it does not have any vertical asymptotes. The $\lim_{x \to \pm\infty} y(x) = \lim_{x \to \pm\infty} (mx + b) = (-\infty)$ or $(+\infty)$ depending on the sign of m, the slope. Therefore there are no horizontal asymptotes, except of course in the case of a horizontal line.

The x-intercept of a line is found by setting y = 0 $\Rightarrow$ 0 = mx + b

$\Rightarrow \quad x = -\frac{b}{m}$ will be the x-intercept.

The y-intercept of a line is found by setting x = 0 $\quad \Rightarrow \quad$ y = m(0) + b
$\Rightarrow$ y = b will be the y-intercept.

Lines do not have any local minimums or maximums.

(**Quadratic Functions**): A quadratic function (or parabola) is a function of the form $y = ax^2 + bx + c$, where a,b, and c are constants. Its domain is all real numbers. We know that a parabola is a curve opening upward with a minimum at a certain point, or a curve opening downward with a maximum at a certain point.

y′(x) = 2ax + b. If we set y′(x) = 0, we will find its only critical value x = $\left(\frac{-b}{2a}\right)$, which is the only x-value where the tangent line will be horizontal and this x-value is where the parabola has a minimum or maximum value. We call this point $\left(\frac{-b}{2a}, y\left(\frac{-b}{2a}\right)\right)$ its vertex. If the parabola opens downward, then y′(x) is positive for all points on the curve to the left of the vertex, and y′(x) is negative for all points on the curve to the right of the vertex. If the parabola opens upward, then y′(x) is negative for all points on the curve to the left of the vertex, and y′(x) is positive for all points on the curve to the right of the vertex.

y″(x) = 2a. This says that the concavity of a parabola is always negative or always positive (depending on the value of a) for all x-values in its domain, the real numbers. If a < 0, then the parabola always bend downward as we go from left to right, so it opens downward. If a > 0, then the parabola always bends upward as we go from left to right, so it opens upward. So if a < 0, there will be a maximum at the vertex $\left(\frac{-b}{2a}, y\left(\frac{-b}{2a}\right)\right)$. If a > 0, there will be a minimum at the vertex $\left(\frac{-b}{2a}, y\left(\frac{-b}{2a}\right)\right)$.

Since y(x) is defined for all x, there are no vertical asymptotes. Since $\lim_{x \to \pm\infty} y(x) = \lim_{x \to \pm\infty} (ax^2 + bx + c) = (-\infty)\ or\ (+\infty)$, depending on the value of a, there are no horizontal asymptotes.

The x-intercepts are found from setting y = 0, or in other words solving the quadratic equation $ax^2 + bx + c = 0$. This quadratic equation will yield no real roots (the parabola doesn't cross the x-axis), or one repeated real root (the parabola's vertex touches the x-axis at only one point), or two distinct real roots (the parabola crosses the x-axis at two distinct points).

The y-intercept is y(0) $\Rightarrow$ y(0) = $a(0)^2 + b(0) + c$ = c .

<u>Example 1</u>: Graph the Parabola y(x) = $x^2 - 2x - 8$.

y'(x) = 2x - 2. Setting this equal to 0, we have x = 1 as the only critical value. This is the x-value of the vertex, and its associated y-value is y(1) = -9. Therefore the vertex is at (1,-9).
Choosing x = 0 as a test point to the left of 1, y'(0) = -2, therefore the curve is decreasing on the interval $(-\infty, 1)$.
Choosing x = 2 as a test point to the right of 1, y'(2) = 2, therefore the curve is increasing on the interval $(1, \infty)$.

y(x) is defined for all x, so there are no vertical asymptotes. The $\lim_{x \to \pm\infty} y(x) = \lim_{x \to \pm\infty} (x^2 - 2x - 8) = (+\infty)$, so there are no horizontal asymptotes.

y"(x) = 2. Since this is positive for all x, the curve is concave up for all x in its domain. So we have a local minimum at the vertex.

The x-intercepts are solutions of $x^2 - 2x - 8 = 0$
$\Rightarrow$ (x - 4)(x + 2) = 0 $\Rightarrow$ x = {-2, 4} .

The y-intercept is y(0) = -8 . From all of this information, we can graph y(x):

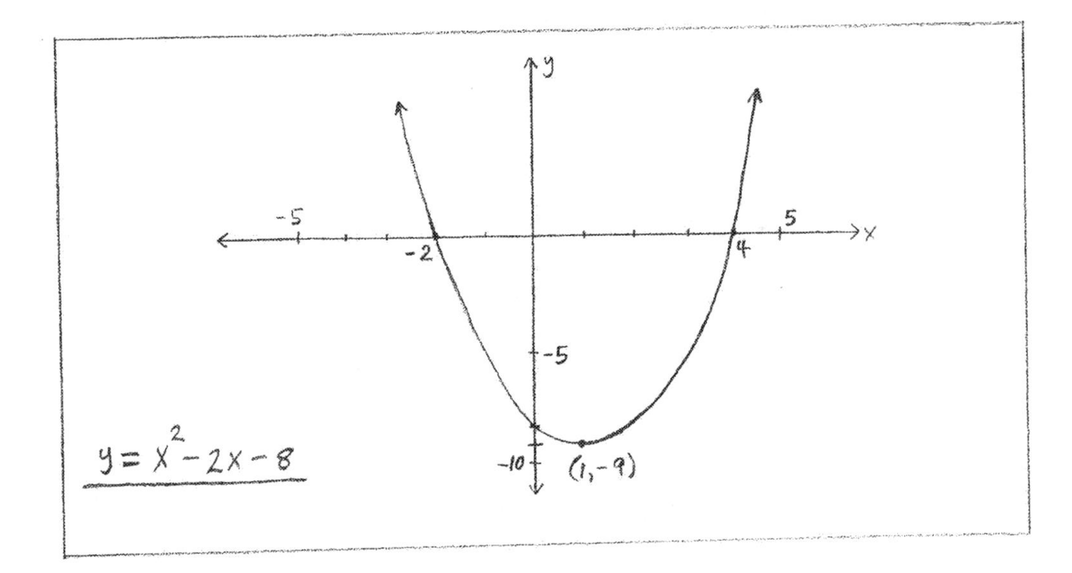

(**Cubic Functions**): A cubic function (or a cubic polynomial function) is of the form $y = ax^3 + bx^2 + cx + d$, where a,b,c, and d are constants. For this function, y'(x) = $3ax^2 + 2bx + c$, and we would find the critical values by solving the corresponding quadratic equation $3ax^2 + 2bx + c$ = 0. The second derivative y''(x) = $6ax + 2b$, and we would find the inflection values by solving the corresponding linear equation $6ax + 2b$ = 0. The critical values would break the x-axis into a collection of intervals, on which we would use test points to determine if y(x) is decreasing, increasing, or constant on each interval. The inflection values would break the x-axis into a collection of intervals, on which we would use test points to determine if y(x) is concave down, concave up, or of zero concavity on each interval. We use the fact that y(x) is continuous for all real numbers x, and where y(x) is increasing or decreasing, along with where y(x) is concave up or down, to determine where there are any local minimums or maximums.

Since this y(x) is defined everywhere, it has no vertical asymptotes. Since the $\lim_{x \to \pm\infty} (ax^3 + bx^2 + cx + d)$ = $(-\infty)$ or $(+\infty)$ depending on the sign of a, there are no horizontal asymptotes. So just as with lines and parabolas, there are no vertical or horizontal asymptotes (except for horizontal lines). In fact there are no vertical or horizontal asymptotes for any kind of polynomial function, regardless of the degree. The so-called end behavior of any polynomial function is to head off for $(-\infty)$ or $(+\infty)$, it may be the same or different in the two directions. As x gets very large negatively or positively, in order to make the determination of whether the end behavior is $(-\infty)$ or $(+\infty)$ depends on the degree of the polynomial function and the coefficient on the highest power term.

The x-intercepts would be the roots of the above cubic function (which can be found by techniques usually learned in pre-calculus or with the aid of a computer). The y-intercept is y(0), which for the cubic function described above is the constant d.

Example 2: Graph the cubic function y(x) = $x^3 - 3x^2 - x + 3$.
y'(x) = $3x^2 - 6x - 1$. Setting this equal to 0, we get the quadratic equation in one variable $3x^2 - 6x - 1 = 0$. From the quadratic formula, we have
x = $\frac{6 \pm \sqrt{36-(4)(3)(-1)}}{6}$ = $\frac{6 \pm \sqrt{48}}{6}$ $= 1 \pm \frac{2\sqrt{3}}{3} \approx \{-0.155,\ 2.155\}$. These are the critical values.

y"(x) = $6x - 6$. Setting this equal to 0, we have the single inflection value {1}.

Plugging these three numbers back into y(x), we have
y(-0.155) $\approx$ 3.08, y(2.155) $\approx$ -3.08, and y(1) = 0. Therefore the critical points are (-0.155, 3.08) and (2.155, -3.08). The inflection point is (1,0).

The roots can be found to be {-1, 1, 3}. Since y(x) is a polynomial it is continuous for all real numbers and there are no vertical or horizontal asymptotes, and since the leading coefficient of y(x) is positive, y(x) goes to $(-\infty)$ as x$\rightarrow (-\infty)$ and y(x) goes to $(+\infty)$ as x$\rightarrow (+\infty)$.
The two critical values break the x-axis into three separate open intervals. We need to choose a test point from each interval to see whether y(x) is increasing or decreasing on that interval.

On $(-\infty, -0.155)$, choose test point -1. y′(-1) = 8, which is positive. Therefore y(x) is increasing on this interval.
On $(-0.155, 2.155)$, choose test point 0. y′(0) = -1, which is negative. Therefore y(x) is decreasing on this interval.
On $(2.155, \infty)$, choose test point 3. y′(3) = 8, which is positive. Therefore y(x) is increasing on this interval.

The single inflection value breaks the x-axis into two separate open intervals. We need to choose a test point from each interval to see whether y(x) is concave up or concave down on that interval.
On $(-\infty, 1)$, choose test point 0. y″(0) = -6, which is negative. Therefore y(x) is concave down on this interval.
On $(1, \infty)$, choose test point 2. y″(2) = 6, which is positive. Therefore y(x) is concave up on this interval.

Using all of this information, we can sketch the graph of y(x):

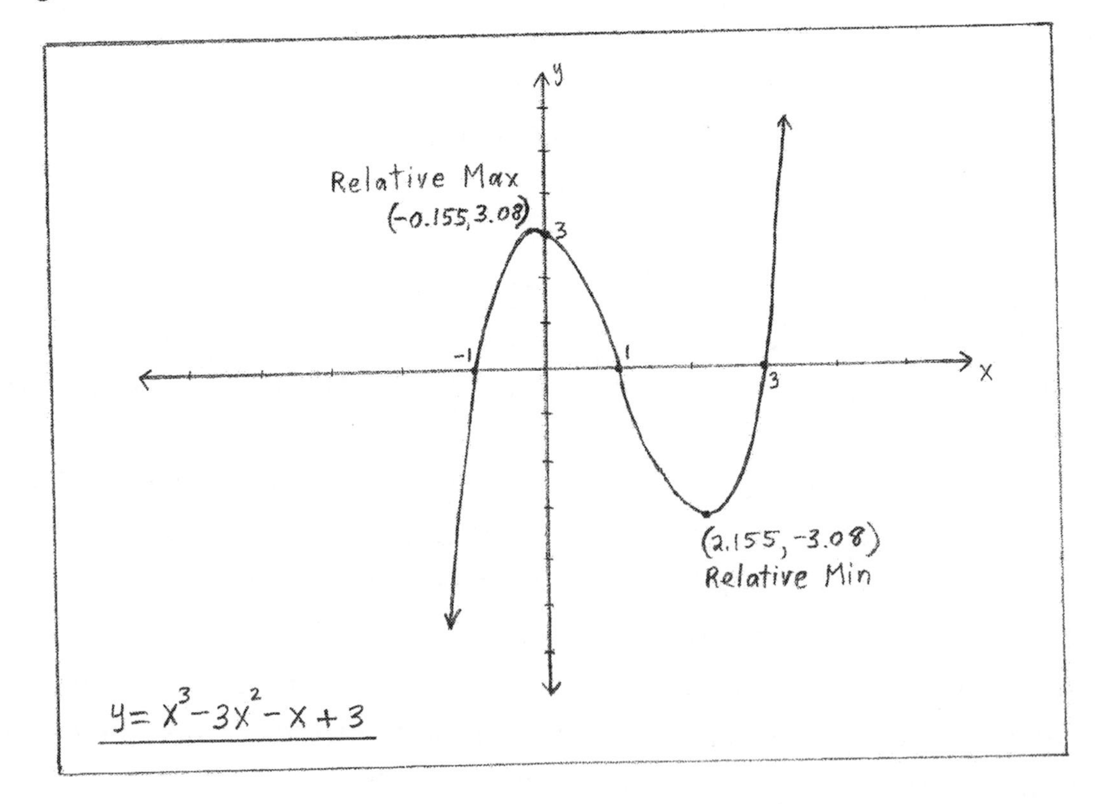

(**Rational Functions**): A rational function is one which is the ratio of two polynomials. The first consideration should be vertical asymptotes. Anywhere that the denominator is zero and the numerator is non-zero, there will be a vertical asymptote and we remove this x-value from the domain. Anywhere that both the numerator and denominator are zero is where the function will have an undefined point without a vertical asymptote (a hole in the graph), so we remove this x-value from the domain.

The next step is to calculate the first and second derivatives. Where y′(x) is 0 or undefined, those are the critical values and we can determine using test points whether the function is increasing or decreasing on each of the intervals determined by them. Where y″(x) is 0 or undefined, those are the inflection values and we can determine using test points whether the function is concave up or down on each of the intervals determined by them.

Then we can examine the $\lim_{x \to -\infty} y(x)$ and the $\lim_{x \to +\infty} y(x)$ to determine any horizontal asymptotes. Then figure if there are any x-intercepts by setting y(x) = 0, and any y-intercepts by determining y(0). From the information determined about where y′(x) = 0 and the concavity at those points we can determine if we have any local minimums or maximums.

Example 3: Analyze and graph y = $\frac{1}{x}$.

The first consideration is the fact that the denominator is 0 at x = 0, so there will be a vertical asymptote at x = 0 and this x-value is not in the domain of y(x).

From the quotient rule, we determine that y′(x) = $\frac{-1}{x^2}$ and y″(x) = $\frac{2}{x^3}$.
y′(x) is always negative for any real number, but is not defined for x = 0. So the set of critical values is {0}.

y''(x) is always negative for x < 0, and always positive for x > 0. It is not defined at x = 0. So the set of inflection values is {0}.

Since y'(x) is always negative, y(x) is decreasing for both intervals $(-\infty, 0)$ and $(0, \infty)$. y(x) is concave down on $(-\infty, 0)$ and concave up on $(0, \infty)$.

The horizontal asymptote is $\lim_{x \to \pm\infty} \frac{1}{x}$ = 0, but y(x) approaches 0 from the negative side (from below) as x goes to $(-\infty)$, and y(x) approaches 0 from the positive side (from above) as x goes to $(+\infty)$.
There are no x or y intercepts. The $\lim_{x \to 0^-} \frac{1}{x} = (-\infty)$. The $\lim_{x \to 0^+} \frac{1}{x} = (+\infty)$.
Using all of this information, we can graph y(x):

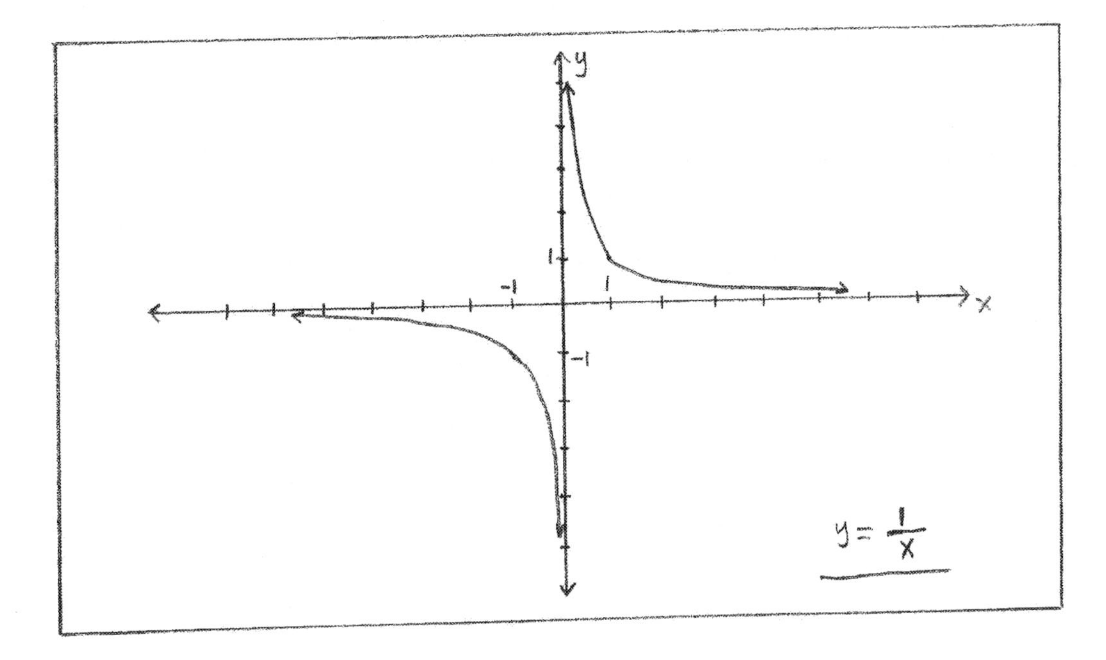

Example 4: Analyze and graph y = $\frac{x^2 - 1}{x^2 - 5x - 6}$.
y = $\frac{x^2 - 1}{x^2 - 5x - 6} = \frac{(x-1)(x+1)}{(x-6)(x+1)} = \frac{(x-1)}{(x-6)}$. The denominator is zero and the numerator non-zero for x = 6, so there is a vertical asymptote at x = 6, and we remove x = 6 from the domain of y(x). The denominator and numerator are both zero at x = -1, and the $\lim_{x \to -1} y(x) = \lim_{x \to -1} \frac{(x-1)}{(x-6)} = \frac{2}{7}$. The function

approaches $\frac{2}{7}$ as x approaches x = -1, but the function is not defined there (there is a hole there) and we remove x = -1 from the domain.

From the quotient rule we can figure y'(x) = $\frac{(-5)(x+1)^2}{(x-6)^2(x+1)^2} = \frac{(-5)}{(x-6)^2}$. This is not defined for x = -1 and for x = 6. Note it is also never 0, so there are no local minimums or maximums. It is always negative where it is defined, so y(x) is always decreasing on its domain. Therefore the critical values are the set of x-values {-1,6}.

From the quotient rule we can figure y"(x) = $\frac{10}{(x-6)^3}$. This is not defined for x = 6, and also it is never 0. Therefore, the inflection values are the set of values {6}.

The horizontal asymptote is $\lim_{x \to \pm\infty} \frac{x^2-1}{x^2-5x-6}$ = 1.
The x-intercept is only x = 1 (x = -1 was removed from the domain)
The y-intercept is y(0) = $\frac{1}{6}$.

The critical values break the domain into the three intervals $(-\infty,-1)$, (-1,6), and (6, ∞). We will use the test values x = -2, 0, and 7 for these intervals respectively.
y'(-2) = $\frac{-5}{64}$ which is negative. Therefore y(x) is decreasing on $(-\infty,-1)$.
y'(0) = $\frac{-5}{36}$ which is negative. Therefore y(x) is decreasing on (-1,6).
y'(7) = -5 which is negative. Therefore y(x) is decreasing on (6, ∞).

The inflection values break the domain into the two intervals $(-\infty,6)$ and $(6,\infty)$. We will use the test values 0 and 7 for these intervals respectively.
y"(0) = $\frac{-5}{108}$ which is negative. Therefore y(x) is concave down on $(-\infty,6)$.
y"(7) = 10 which is positive. Therefore y(x) is concave up on (6, ∞).
We can put all this information together to graph y(x):

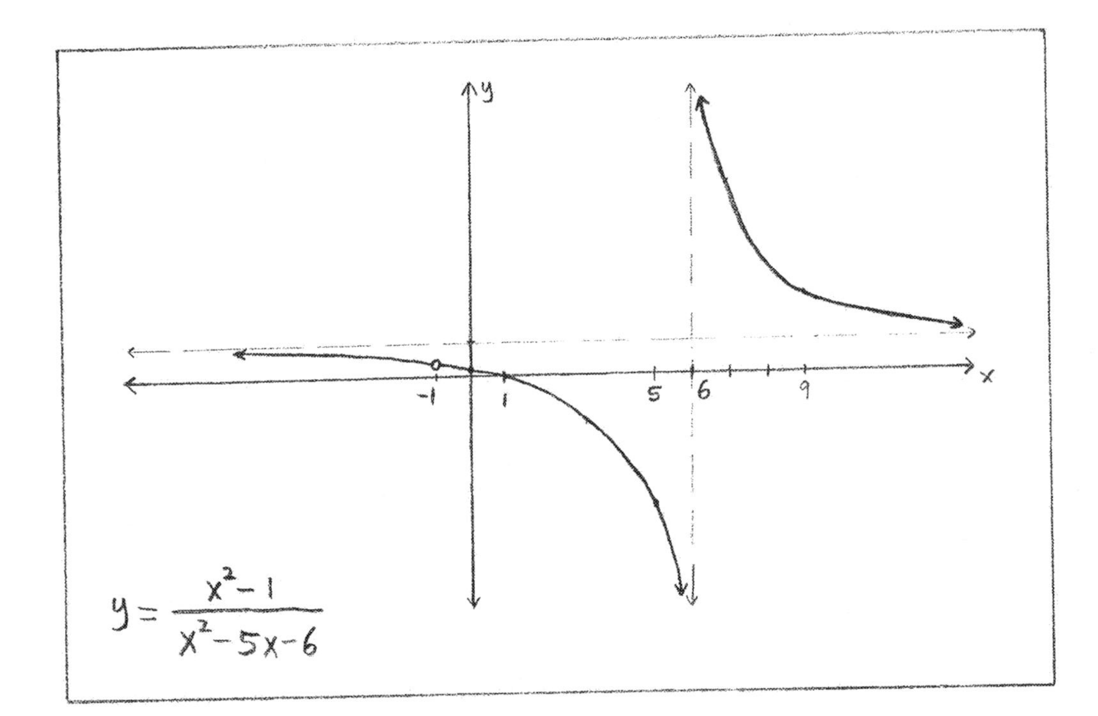

(**The Square Root and Cube Root Functions**):

(A) The Square Root Function.

This is the function y = $\sqrt{x} = x^{\frac{1}{2}}$. Its domain is x $\geq$ 0.

y'(x) = $\frac{1}{2}x^{-\frac{1}{2}} = \frac{1}{2\sqrt{x}}$. For x > 0, y'(x) is always positive, which means that y(x) is always increasing. y'(x) is never 0, which means that y(x) has no local minimums or maximums. Note that x = 0 is the only critical value because y'(x) isn't defined there. However, the $\lim\limits_{x \to 0^+} y'(x) = (+\infty)$.

y"(x) = $\left(-\frac{1}{4}\right)\left(x^{\frac{-3}{2}}\right) = -\frac{1}{4}\left(\frac{1}{(\sqrt{x})^3}\right)$. For x > 0, this is always negative, which means that y(x) is always concave down.

Since it is defined for all x $\geq$ 0, y = $\sqrt{x}$ has no vertical asymptotes. The $\lim\limits_{x \to \infty} \sqrt{x} = (+\infty)$, so it has no horizontal asymptotes.

Its x-intercept is x = 0, and its y-intercept is y = 0. This function is graphed back in the chapter where we graphed common functions.

(B) The Cube Root Function.

This is the function y = $\sqrt[3]{x} = x^{\frac{1}{3}}$, which is defined for all real numbers.

y'(x) = $\frac{1}{3}x^{-\frac{2}{3}} = \frac{1}{3(\sqrt[3]{x})^2}$, which is always positive. Therefore the cube root function is always increasing. y'(x) is never 0, so it has no local minimums or maximums. Note that x = 0 is the only critical value, since y'(x) is not defined there. However, the $\lim_{x \to 0} y'(x) = (+\infty)$.

y''(x) = $(-\frac{2}{9})x^{-\frac{5}{3}} = (-\frac{2}{9})\frac{1}{(\sqrt[3]{x})^5}$. x = 0 is the only inflection value since y''(x) is not defined there. For x < 0, y''(x) is positive. Therefore, for x < 0, y(x) is concave up. For x > 0, y''(x) is negative. Therefore, for x > 0, y(x) is concave down.

Since y(x) = $\sqrt[3]{x}$ is defined for all real numbers, it has no vertical asymptotes. Since the $\lim_{x \to -\infty} \sqrt[3]{x} = (-\infty)$, and the $\lim_{x \to \infty} \sqrt[3]{x} = (+\infty)$, y(x) has no horizontal asymptotes.

The x-intercept and the y-intercept are both 0. This function is graphed back in the chapter where we graphed common functions.

(**The Natural Logarithmic and Exponential Functions**):

(A) The natural logarithmic function is y = ln(x), and has domain x > 0.

$y'(x) = \frac{1}{x}$, is always positive for x > 0. Therefore y(x) is a monotone increasing function, with no local minimums or maximums. x = 0 is its only critical value because y'(x) is not defined there.

$y''(x) = \frac{-1}{x^2}$, which is always negative for x > 0. Therefore y(x) = ln(x) is concave down on its domain. x = 0 is the only inflection value since y''(x) is undefined there.

y = ln(x) has x = 0 (the y-axis) as its only vertical asymptote. The $\lim_{x \to 0^+} ln(x) = (-\infty)$. The $\lim_{x \to \infty} ln(x) = (+\infty)$. So it has no horizontal asymptote.

The x-intercept is x = 1 (ln(1) = 0). There is no y-intercept. This function is graphed back in the chapter where we graphed common functions.

(B) The natural exponential function is y = e^x, which is defined for all real numbers x, and it is positively valued for all x.

y'(x) = e^x is positive for all x. Therefore y = e^x is a monotone increasing function, with no local minimums or maximums on its entire domain.

y''(x) = e^x is positive for all x. Therefore y = e^x is concave up on its entire domain.

y = e^x has no vertical asymptotes. However, the $\lim_{x \to -\infty} e^x$ = 0 and the $\lim_{x \to \infty} e^x = \infty$. Therefore, y = e^x has no asymptote as x goes to ($+\infty$), but it has y = 0 (the x-axis) as a horizontal asymptote as x goes to ($-\infty$), and it approaches this horizontal asymptote from above.

y = e^x has no x-intercept, but its y-intercept is 1 ($e^0 = 1$). This function is graphed back in the chapter where we graphed common functions.

EXERCISES:

For each of the following functions y(x) below, state:

(a) any vertical asymptotes

(b) any horizontal asymptotes

(c) y′(x) and critical values, y″(x) and inflection values

(d) intervals where y(x) is increasing

(e) intervals where y(x) is decreasing

(f) intervals where y(x) is concave up

(g) intervals where y(x) is concave down

(h) any local minimums or maximums

(i) any absolute minimums or maximums

(j) x-intercepts and y-intercepts

(1.) $y = \frac{1}{x^2}$

(2.) $y = \sin(x)$ (for: $0 < x < 2\pi$)

(3.) $y = \tan(x)$ (for: $-\frac{\pi}{2} < x < \frac{\pi}{2}$)

(4.) $y = e^{-x}$

(#) **Optimization Problems**:

An optimization problem is one where we have a function to be minimized or maximized and we use the first derivative and its critical values to find potential solutions, and then use the second derivative to determine if indeed we have found an optimal solution. The function to be minimized or maximized will be continuous on some open interval which contains the solution.

Example 1: A box with a square base must have a volume of 500 ft^3. What dimensions of the box will minimize the surface area?

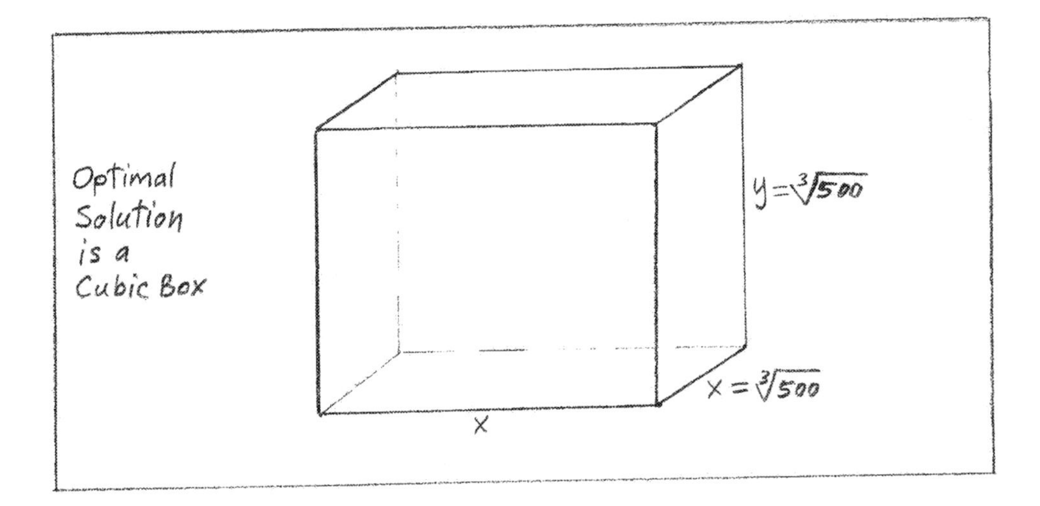

Let x = the length of the sides of the base, and let y = the height of the box. Then we want to minimize the surface area $S = 2x^2 + 4xy$ subject to the constraint that the volume $V = x^2y$ must be 500 ft^3. From the constraint, we see that $y = \frac{500}{x^2}$. Substituting in the function for y in order to get a function of the one variable x, we have $S = 2x^2 + \frac{2000}{x}$. S is to be minimized. By considering the extremes that x can take given our constraint, we will need the optimal x to be in the interval $(0, \sqrt{500})$. Now we use the first and second derivatives to find the minimum for S.

$\frac{dS}{dx} = 4x - \frac{2000}{x^2}$, which we set equal to 0 to find any critical values in the interval $(0, \sqrt{500})$. $4x - \frac{2000}{x^2} = 0 \quad \Rightarrow \quad 4x = \frac{2000}{x^2} \quad \Rightarrow \quad x^3 = 500$

$\Rightarrow \quad x = \sqrt[3]{500}$, which is in $(0, \sqrt{500})$.

So, $y = \frac{500}{x^2} = \frac{500}{(\sqrt[3]{500})^2} = \sqrt[3]{500}$ also. We should use the second derivative to make sure that these values actually minimize the surface area S.

$\frac{d^2S}{dx^2} = 4 + \frac{4000}{x^3}$. Evaluating this at $\sqrt[3]{500}$, leads to $4 + \frac{4000}{500} = 12$. Since this value for the second derivative is positive, then we know that we have indeed found a minimum. The actual minimum surface area is

$S = 2(\sqrt[3]{500})^2 + 4(\sqrt[3]{500})^2 = 6(\sqrt[3]{500})^2 \approx 378\ ft^2$.

Therefore, the optimal solution is a cubical box where all three sides are of length ($\sqrt[3]{500}$) ft, which is approximately 7.937 ft. This is a feasible solution since $\sqrt[3]{500} \in (0, \sqrt{500})$.

Example 2: Find the rectangle of maximum area that can be inscribed within the ellipse $\frac{x^2}{a^2} + \frac{y^2}{b^2} = 1$.

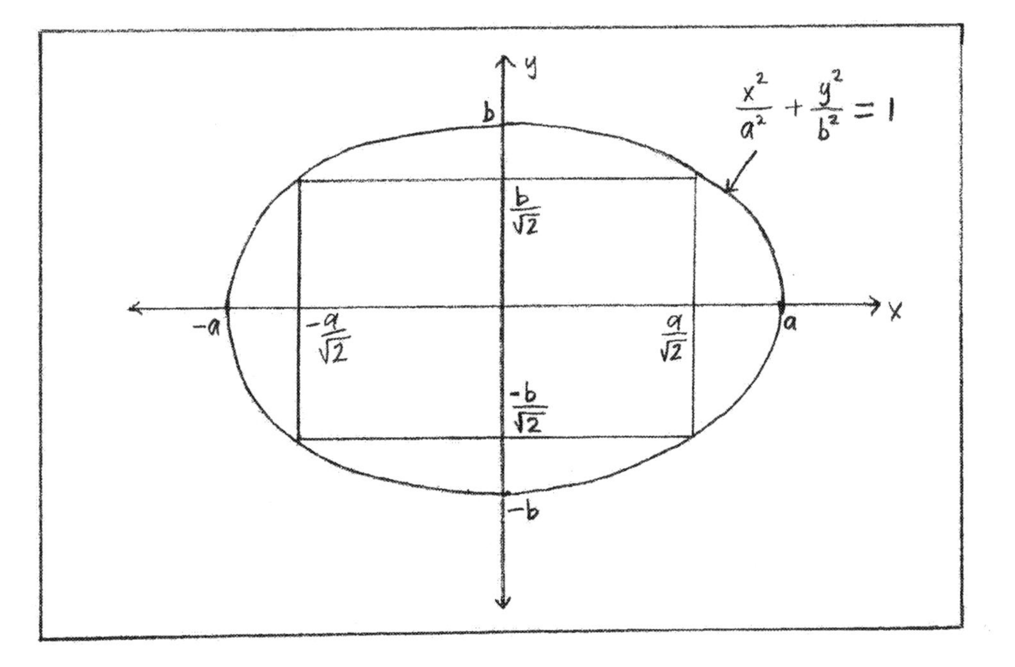

We can restrict ourselves to the first quadrant, where $0 < x < a$, $0 < y < b$. Solving for y, we have

$\frac{y^2}{b^2} = 1 - \frac{x^2}{a^2} \quad \Rightarrow \quad y = b\left(1 - \frac{x^2}{a^2}\right)^{\frac{1}{2}}$. We want to maximize the Area $A_1 = xy$, for $x \in (0, a)$.

$A_1 = xy = bx\left(1 - \frac{x^2}{a^2}\right)^{\frac{1}{2}}$ so that

$\frac{dA_1}{dx} = (bx)(\frac{1}{2})\left(1 - \frac{x^2}{a^2}\right)^{-\frac{1}{2}}(\frac{-2x}{a^2}) + (b)\left(1 - \frac{x^2}{a^2}\right)^{\frac{1}{2}}$ which we set equal to 0 in order to find any critical values:

$\Rightarrow \quad \left(1 - \frac{x^2}{a^2}\right)^{\frac{1}{2}} = (\frac{x^2}{a^2})\left(1 - \frac{x^2}{a^2}\right)^{-\frac{1}{2}} \quad \Rightarrow \quad \left(1 - \frac{x^2}{a^2}\right) = \left(\frac{x^2}{a^2}\right)$

$\Rightarrow \quad 1 = \frac{2x^2}{a^2} \quad \Rightarrow \quad x = \frac{a}{\sqrt{2}} \quad \Rightarrow \quad y = b\left(1 - \frac{x^2}{a^2}\right)^{\frac{1}{2}} = \frac{b}{\sqrt{2}}$. This critical value $x = \frac{a}{\sqrt{2}}$, is in the interval $(0,a)$.

So the rectangle of maximum area will have dimensions $L = 2x = \sqrt{2}a$ and $H = 2y = \sqrt{2}b$, since we found the optimal x and y restricting ourselves to the first quadrant. The area of this rectangle is $A = 4A_1 = 2ab$. We want to make sure that this solution is a maximum, so we will substitute our optimal x-value into the second derivative:

$\frac{d^2A_1}{dx^2} = (\frac{b}{2})\left(1 - \frac{x^2}{a^2}\right)^{-\frac{1}{2}}(\frac{-2x}{a^2}) - \left[(\frac{bx^2}{a^2})(-\frac{1}{2})\left(1 - \frac{x^2}{a^2}\right)^{-\frac{3}{2}}(\frac{-2x}{a^2}) + \left(1 - \frac{x^2}{a^2}\right)^{-\frac{1}{2}}(\frac{2bx}{a^2})\right]$. This second derivative evaluated at $x = \frac{a}{\sqrt{2}}$ is $((-\frac{b}{a}) - (\frac{3b}{a})) = \frac{-4b}{a} < 0$.
Since this is negative, we do indeed have a maximum.

Example 3: A farmer must fence off a rectangular region alongside a river with 5300 yds of fence. So the river will be one of the sides of the rectangle and will not require fencing. If we call the side parallel to the river x and the other two sides y, then we have the constraint
$L = x + 2y = 5300$. Find the values for x and y that will maximize the area A enclosed by the fence.

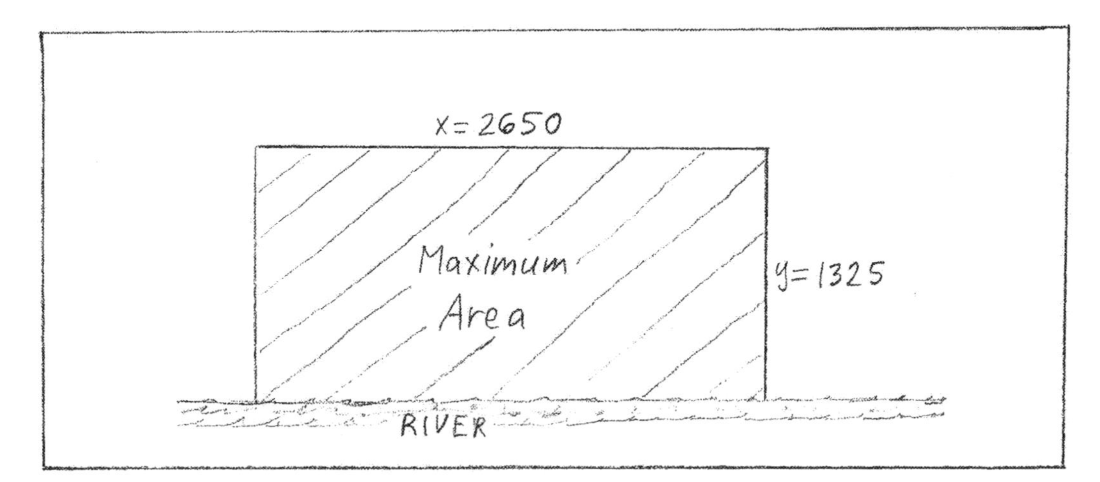

A = xy, where x $\in (0, 5300)$.

The constraint x + 2y = 5300 $\Rightarrow$ y = $\frac{1}{2}(5300 - x) = 2650 - \frac{x}{2}$.

So A = xy = x$(2650 - \frac{x}{2}) = 2650x - \frac{x^2}{2}$ is a function of one variable.

Then $\frac{dA}{dx} = (2650 - x)$ which we set equal to 0 to find the critical values.

$\Rightarrow$ x = 2650 is the critical value. Then this critical value x and

y = $(2650 - \frac{x}{2}) = (2650 - \frac{2650}{2})$ = 1325 are the solutions we seek.

$\frac{d^2A}{dx^2} = (-1)$, which is of course negative for all x. Therefore, our solution is a maximum. The maximum area enclosed is equal to (2650)·(1325) = 3,511,250 yd^2.

<u>Example 4</u>: Suppose we have a real line X $(-\infty < x < \infty)$ with the origin at point O. The point P is a distance of 5 units from the real line X, with line segment OP perpendicular to X. See the figure below. For x $\in (-\infty, \infty)$, let θ be the angle between the line containing O and P, and the line through P and the point x on the line X. Find where $\frac{d\theta}{dx}$ is a maximum, as x goes from $-\infty$ to $+\infty$. From the diagram below,

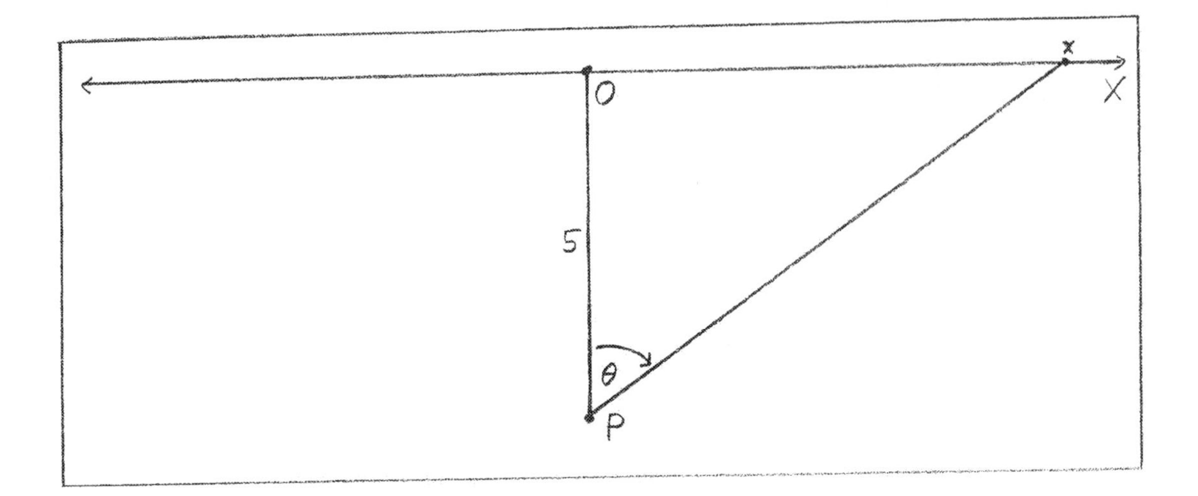

we can see that $\theta = tan^{-1}\left(\frac{x}{5}\right)$. So $\frac{d\theta}{dx}$ = $\left(\frac{1}{1+\frac{x^2}{25}}\right)\cdot\left(\frac{1}{5}\right)$. We can see that this is maximized when x = 0, but we want to use calculus to show this. The critical values for $\frac{d\theta}{dx}$ are found by setting the second derivative $\frac{d^2\theta}{dx^2}$ equal to 0. So, $\frac{d^2\theta}{dx^2}$ = $\left(\frac{1}{5}\right)\left((-1)\frac{2x}{25}\right)\left(1+\frac{x^2}{25}\right)^{-2}$ = $\left(\frac{-2}{125}\right)\left(\frac{x}{(1+\frac{x^2}{25})^2}\right)$, which we set = 0. It is easy to see that x = 0 is the only critical value. To the left of x = 0, $\frac{d^2\theta}{dx^2}$ is always positive. To the right of x = 0, $\frac{d^2\theta}{dx^2}$ is always negative. This second derivative is continuous for all real numbers x. Therefore, we can see that the rate of change of θ with respect to x, $\frac{d\theta}{dx}$, is maximized at x = 0.

Example 5: Physicists are studying a new type of particle with never before seen properties, that they call a newon. They believe that it is repelled by a single electron's negative charge at a rate that varies according to the proximity of the newon and the electron. The physicists are modeling the repulsive force (in Newtons) on the newon as F(x) = 20x + $\frac{10}{x^2}$, where x is in meters and x > 0. So strangely enough the newon is strongly repelled when very close and very far from electrons, but there is an intermediate distance at which the repulsion is at a minimum. What is the minimum repulsive force experienced by the newon and at what distance x in meters?

F'(x) = $20 - \frac{20}{x^3}$, which we set equal to zero to find the critical value. So, $\frac{20}{x^3} = 20$ $\Rightarrow$ x = 1 is the critical value and falls in the region $(0, \infty)$. F"(x) = $\frac{60}{x^4}$, which is always positive, so our critical value corresponds to a minimum. So the physicists conclude that the repulsion F(x) is minimized at a distance of exactly 1 meter, and it is 30 Newtons of force!

(#) <u>Related Rates</u>: Sometimes we start with a relation between a collection of variables, and we are able to derive an equation (through implicit differentiation with respect to time) that expresses the relationship between each of the variables, along with their derivatives with respect to time. So therefore we are assuming that each of the variables is an implicit function of time t. If at an instant in time, we know the values of the variables and all but one of the rates of change, we are able to solve for the remaining one.

<u>Example 1</u>: If we have a right triangle with legs x and y, and hypotenuse z, then we know from the pythagorean theorem that $x^2 + y^2 = z^2$. If at the instant that x = 7 and y = 9, x is increasing at the rate $\frac{dx}{dt}$ = 2 ft./second, and y is changing at the rate $\frac{dy}{dt}$ = -0.8 ft./second, how fast is z changing?

From $x^2 + y^2 = z^2$, after differentiating all of the variables with respect to time, we have $2x\frac{dx}{dt}$ + 2y$\frac{dy}{dt}$ = 2z$\frac{dz}{dt}$ or $\frac{dz}{dt}$ = $\frac{x\frac{dx}{dt} + y\frac{dy}{dt}}{z}$. This equation relates x, y, z, $\frac{dx}{dt}$, $\frac{dy}{dt}$, and $\frac{dz}{dt}$. Now, when x = 7 and y = 9, then z = $\sqrt{49+81} = \sqrt{130}$. So we can solve for $\frac{dz}{dt}$, the rate of change of the hypotenuse at the specific time in question, from the equation above.

So, z is changing at the rate $\frac{dz}{dt} = \frac{(7)(2)+(9)(-0.8)}{\sqrt{130}} \approx$ 0.5964 ft./second. We can see that z is increasing, at the specific point in time, despite the fact that y is decreasing, since x is increasing much faster.

Example 2: Suppose we have a relation between three variables s, r, and θ given by s = rθ (θ is an angle in a polar coordinate system measured in radians, length from the origin r and arc length s are measured in meters). Assume that r and θ vary with respect to time as $\theta \geq 0$ is increasing, hence s varies with respect to time also. If we know that at a certain point in time s = 25 meters when θ is 2.35 radians, and that the rate of change of r with respect to time is $\frac{dr}{dt}$ = 10.6 meters/sec and that the rate of change of θ with respect to time is $\frac{d\theta}{dt}$ = 1.22 radians/sec , how fast is s changing at that instant?

At the given point in time, $r = \frac{s}{\theta} = \frac{25}{2.35} = 10.638$ meters. From the product rule, $\frac{ds}{dt} = (r)(\frac{d\theta}{dt}) + (\theta)(\frac{dr}{dt})$. So at the time in question, s is changing at the rate $\frac{ds}{dt}$ = (10.638)(1.22) + (2.35)(10.6) = 37.888 meters/sec.

Example 3: A circular track on a large flat surface, where a drawing of an x-y coordinate has been drawn on the surface, is described by the relation $x^2 + y^2 = 100$, where x and y are in feet. A small object moves clockwise along the track with a constant speed. So the speed in the x and y directions varies with respect to time. At the instant that x = -8.2 feet and the object is in the third quadrant, the x-component of its motion is -4.5 ft/sec. What is the y-component of the object's motion?

From $x^2 + y^2 = 100$, differentiating with respect to time, $2x\frac{dx}{dt} + 2y\frac{dy}{dt} = 0$. If x = -8.2 ft., then y = -5.72 ft, because the object is in the third quadrant.

So the y-component of the motion is,

$\frac{dy}{dt} = -\frac{x}{y} \cdot \frac{dx}{dt} = -(\frac{-8.2\,ft}{-5.72\,ft})(-4.5\,ft/sec) = (6.45)$ ft/sec.

EXERCISES:

Solve the following optimization and related rates problems.

(1.) Find the maximum value of S = $9 - x^2 - y^2$ subject to the constraint that y $= 1 - x$, for points (x,y) in the following region of the x-y plane: {(x,y)| $x^2 + y^2 \leq 9$}.

(2.) Find two real numbers x and y from the interval [0, 136] that add to 136, and such that P = xy is maximized. What is the maximum value of P?

(3.) A right circular cylinder must have a volume of 1200 ft^3. Find the base radius r and the height h that minimizes the surface area. What is the minimum surface area?

(4.) Gravel is being dropped into a conical pile with volume V = $\frac{1}{3}\pi r^2 h$. As the gravel accumulates at the instant that r = 10 ft, and h = 8.2 ft, the radius of the pile is changing at a rate $\frac{dr}{dt}$ = 2 ft./minute and the height is increasing at a rate $\frac{dh}{dt}$ = 1.1 ft./minute. How fast is the volume of the pile changing at this point in time?

Part III

The Integral Calculus

(8) Indefinite Integrals

(#) Antiderivatives: If Y(x) and y(x) are two functions such that Y′(x) = y(x), then we say that Y(x) is an antiderivative of y(x).

Example 1: If we have y(x) = 2x, then Y(x) = $x^2 + C$ (for any constant C) is an antiderivative of y(x) since Y′(x) = y(x).

If C = 2, it would be true that Y(x) = $x^2 + 2$ is an antiderivative of y(x) since Y′(x) = $\frac{d}{dx}(x^2 + 2)$ = 2x + 0 = 2x = y(x).

If C = 45, it would be true that Y(x) = $x^2 + 45$ is an antiderivative of y(x) since Y′(x) = $\frac{d}{dx}(x^2 + 45)$ = 2x + 0 = 2x = y(x).

We call C an arbitrary constant. The function y(x) = 2x has an infinite number of antiderivatives, but all of its antiderivatives differ at most by a constant.

Example 2: If we have y(x) = $5x^2 + 8x$, then Y(x) = $\frac{5x^3}{3} + 4x^2 + C$ (for any constant C) is an antiderivative of y(x), since Y′(x) = $\frac{d}{dx}(\frac{5x^3}{3} + 4x^2 + C)$ = $5x^2 + 8x$ = y(x).

(#) Indefinite Integrals: We have a special notation for antiderivatives. If Y(x) and y(x) are two functions and Y′(x) = y(x), we write Y(x) = $\int y(x)dx$ to denote that Y(x) is an antiderivative of y(x). The elongated symbol "$\int$" is called an integral sign and y(x) is called the integrand. We call $\int y(x)dx$ the indefinite integral, or just the integral, of y(x). The "dx" part tells us the variable that we are integrating with respect to. There is another meaning for "dx" which is more natural, that we will see when we consider definite integrals in a later chapter.

Example 1: $\int(10x - e^x)dx = 5x^2 - e^x + C$ is an antiderivative of y(x) = $(10x - e^x)$, since the $\frac{d}{dx}(5x^2 - e^x + C) = (10x - e^x)$ for any constant C.

Example 2: $\int(2)cos(2x)dx = sin(2x) + C$, is an antiderivative of y(x) = (2)cos(2x), since the $\frac{d}{dx}(sin(2x) + C) = (2)cos(2x)$ for any constant C.

(#) Integral Formulas:

Example 1:

If y(x) = x^n, where n ≠ -1, then $\int y(x)dx = \int x^n dx = (\frac{x^{n+1}}{n+1} + C)$,

since $\frac{d}{dx}\left(\frac{x^{n+1}}{n+1} + C\right) = \frac{(n+1)x^n}{(n+1)} + 0 = x^n$.

This is one of our most basic integral formulas, much like the power rule is one of the most basic differentiation formulas. Note that we have to exclude the case n = -1, because we would have $\int x^{-1}dx = (\frac{x^0}{0} + C)$, which is a problem because we can't divide by 0. There is a formula for this case which we will consider soon.

Example 2: If y(x) = a, for any real number a,

The $\int a\,dx$ = (ax + C),

since $\int a\,dx = \int a \cdot x^0 \cdot dx = \frac{ax^{(0+1)}}{(0+1)} + C = (ax + C)$.

The special case where a = 1 is $\int dx = x + C$,

since $\int dx = \int (1)dx$ and $\frac{d}{dx}((1)x + C) = 1$.

We know that for a function y(x), $\frac{\Delta y}{\Delta x} \approx y'(x)$, where Δx, Δy and y'(x) are defined for the point x and another x-value close to x. So $\Delta y \approx y'(x)\Delta x$. Then as $\Delta x \to 0$, this expression is transformed into dy = y'(x)dx. The differential dx always equals the increment Δx, because these are always the same change in the independent variable x. But for the dependent variable y, the increment Δy and the differential dy are generally not equal, they approach each other in the limit as $\Delta x \to 0$. The fact that the differential dy $\approx \Delta y$ as $\Delta x \to 0$ is the basis of the linear approximation that we have talked about before in several examples. The following illustration shows the relationship between these quantities.

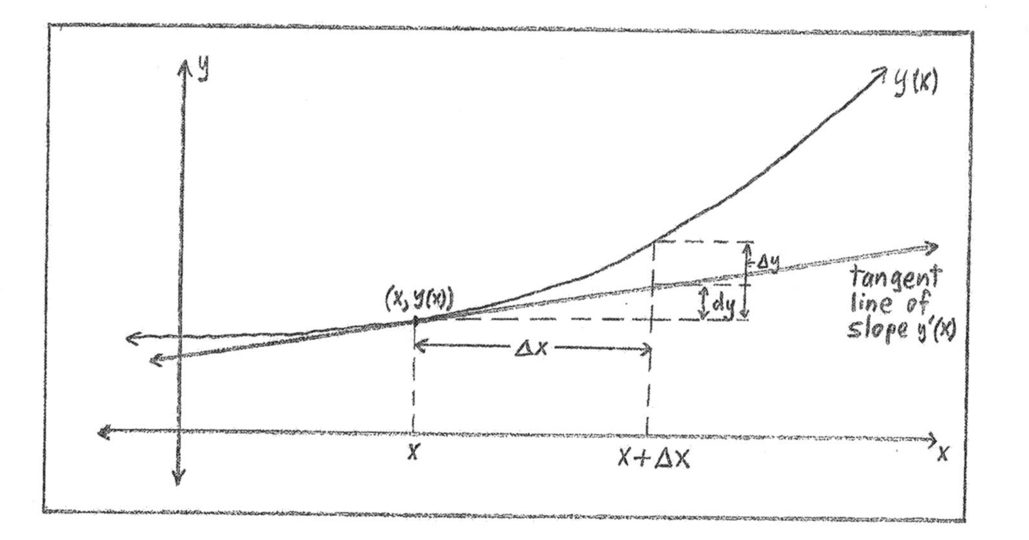

We have a generalization of the above idea. If we have a function of x, call it Q(x), then $\int dQ = Q(x) + C$, because $\int dQ = \int \left(\frac{dQ}{dx}\right) dx = Q(x) + C$.

So we can say: $\int dQ$ = $\int \frac{dQ}{dx} dx = Q(x) + C$, and $\frac{d}{dx} \int Q \, dx = Q(x)$.

<u>Example 3</u>: Now, if the exponents n and m are both not equal to -1,

$\int (ax^n + bx^m) dx \; = \; \int ax^n dx + \int bx^m dx \; = \; a\int x^n dx + b\int x^m dx$

= $[(a)\frac{x^{n+1}}{n+1} + (b)\frac{x^{m+1}}{m+1} + C]$, since $\frac{d}{dx}[(a)\frac{x^{n+1}}{n+1} + (b)\frac{x^{m+1}}{m+1} + C]$ = $(a\,x^n + bx^m)$.

This says that integration satisfies a linearity property just as differentiation does. In other words the integral of a linear combination of functions is the linear combination of the integrals of those functions.

Note that constants can be brought across integral signs.

For example, $\int 7x^3 dx = \left(7\frac{x^4}{4} + C\right) = 7\left(\left(\frac{x^4}{4}\right) + \frac{C}{7}\right) = 7\left(\frac{x^4}{4} + C\right) = 7\int x^3 dx$.

This is true since the constant "$(\frac{C}{7})$" can just as well be called "C".

<u>Example 4</u>: Evaluate $\int (9x + 3x^2 + 4x^4 - 2) dx$.

Using our formulas from above,

$\int (9x + 3x^2 + 4x^4 - 2) dx$ = $\int 9x dx + \int 3x^2 dx + \int 4x^4 dx - \int 2dx$

$= 9\int x dx + 3\int x^2 dx + 4\int x^4 dx - 2\int dx$ = $9\frac{x^2}{2} + 3\frac{x^3}{3} + 4\frac{x^5}{5} - 2x + C$

= $\frac{9}{2}x^2 + x^3 + \frac{4}{5}x^5 - 2x + C$.

We know that this is correct because

$\frac{d}{dx}\left(\frac{9}{2}x^2 + x^3 + \frac{4}{5}x^5 - 2x + C\right)$ = $(9x + 3x^2 + 4x^4 - 2)$.

Note that technically each integration produces its own arbitrary constant, but they can all be lumped together at the end and called C.

Example 5: Find $\int\left(\frac{2x^4 - 6\sqrt{x}}{\sqrt{x}}\right) dx$.

$$\int\left(\frac{2x^4 - 6\sqrt{x}}{\sqrt{x}}\right) dx = \int\left(2x^{\frac{7}{2}} - 6\right) dx = 2\int x^{\frac{7}{2}}dx - 6\int dx$$

$$= \frac{4x^{\frac{9}{2}}}{9} - 6x + C.$$

Example 6: Find $\int(8\sqrt{x} - 4x^{-7})dx$.

$$\int(8\sqrt{x} - 4x^{-7})dx = 8\int x^{\frac{1}{2}}dx - 4\int x^{-7}dx = \frac{16x^{\frac{3}{2}}}{3} + \frac{2}{3}x^{-6} + C$$

$$= \frac{16x^{\frac{3}{2}}}{3} + \frac{2}{3x^6} + C.$$

Example 7: We can use our basic derivative formulas from the differential calculus chapters to develop many integration formulas. We will not derive all of the basic integration formulas, but rather show how certain important ones can be derived from derivative formulas which we already know. The reader can then, as an exercise, verify the remaining ones. However, we will provide a summary of the basic integration formulas at the end of this example.

(A) Since $\frac{d}{dx}(e^x) = e^x$, we have $d(e^x) = e^x dx$, from which we can take the integral of both sides, or as we sometimes say, integrate both sides (with respect to x) to get:

$\int d(e^x) = \int e^x dx$ or $\int e^x dx = e^x + C$.

(B) Since $\frac{d}{dx}(lnx) = \frac{1}{x}$, we have $d(lnx) = \frac{1}{x}dx$, from which

$\int d(lnx) = \int \frac{1}{x}dx$ or $\int \frac{1}{x}dx = ln|x| + C$. (Note that this formula takes care of the problem that we had above in evaluating $\int x^n dx$ when n = -1.)

The reader may be asking why the natural logarithm of the absolute value of x ? Let's consider cases:

Case 1. x > 0. The $\frac{d}{dx}(ln|x| + C)$ = $\frac{d}{dx}(ln(x) + C)$ = $\frac{1}{x}$.
Case 2. x < 0. The $\frac{d}{dx}(ln|x| + C)$ = $\frac{d}{dx}(ln(-x) + C)$ = $(\frac{1}{-x}) \cdot (-1)$ = $\frac{1}{x}$.
So the formula above is correct in all cases.

(C) Since $\frac{d}{dx}(cos(x)) = -sin(x)$, we have $d(cos(x)) = -sin(x)dx$, from which we have $\int d(cos(x)) = -\int sin(x)dx$, or

$\int sin(x)dx = -cos(x) + C$.

(D) Since $\frac{d}{dx}(sin(x)) = cos(x)$, we have d(sin(x)) = cos(x)dx, from which we have $\int d(sin(x)) = \int cos(x)dx$, or

$\int cos(x)dx = sin(x) + C$

.

(E) Since $\frac{d}{dx}(tan(x)) = sec^2(x)$, we have d(tan(x)) = $sec^2(x)dx$, from which we have $\int d(tan(x)) = \int sec^2(x)dx$, or

$\int sec^2(x)dx = tan(x) + C$

.

(F) Since $\frac{d}{dx}(sin^{-1}(x)) = \frac{1}{\sqrt{1-x^2}}$, we have $d(sin^{-1}(x)) = \frac{1}{\sqrt{1-x^2}}\, dx$, from which we have $\int d(sin^{-1}(x)) = \int \frac{1}{\sqrt{1-x^2}}\, dx$, or

$\int \frac{1}{\sqrt{1-x^2}}\, dx = sin^{-1}(x) + C$.

(G) Since $\frac{d}{dx}(tan^{-1}(x)) = \frac{1}{1+x^2}$, we have $d(tan^{-1}(x)) = \frac{1}{1+x^2}\, dx$, from which we have $\int d(tan^{-1}(x)) = \int \frac{1}{1+x^2}\, dx$, or

$\int \frac{1}{1+x^2}\, dx = tan^{-1}(x) + C$.

(H) Since $\frac{d}{dx}(sec^{-1}(x)) = \frac{1}{x\sqrt{x^2-1}}$, we have $d(sec^{-1}(x)) = \frac{1}{x\sqrt{x^2-1}}\, dx$, from which we have $\int d(sec^{-1}(x)) = \int \frac{1}{x\sqrt{x^2-1}}\, dx$, or

$\int \frac{1}{x\sqrt{x^2-1}}\, dx = sec^{-1}(x) + C$.

(**Summary of Basic Integration Formulas**)

$\int x^n dx = \frac{x^{n+1}}{n+1} + C$, (n ≠ -1)

$\int e^x dx = e^x + C$

$\int \left(\frac{1}{x}\right) dx = ln|x| + C$

$\int a^x dx = \frac{a^x}{ln(a)} + C$, a ≠ e.

$\int \left(\frac{1}{x(lna)}\right) dx = log_a(x) + C$, a ≠ e.

$\int sin(x)\, dx = -cos(x) + C$

$\int cos(x)\, dx = sin(x) + C$

$\int tan(x) = -ln|cos(x)| + C$

$\int cot(x)\, dx = ln|sin(x)| + C$

$\int sec(x)\, dx = ln|sec(x) + tan(x)| + C$

$\int csc(x)\, dx = -ln|csc(x) + cot(x)| + C$

$\int sec^2(x)\, dx = tan(x) + C$

$\int csc^2(x)\, dx = -cot(x) + C$

$\int sec(x)tan(x)\, dx = sec(x) + C$

$\int csc(x)cot(x)\, dx = -csc(x) + C$

$\int \frac{1}{\sqrt{1-x^2}}\, dx = sin^{-1}(x) + C$

$\int \frac{1}{1+x^2}\, dx = tan^{-1}(x) + C$

$\int \frac{1}{x\sqrt{x^2-1}}\, dx = sec^{-1}(x) + C$

We are now at a point where we can discuss a few important techniques for finding the integral of more complicated functions. It should be noted that unlike differentiation, where it is always possible to determine the derivative of any continuous function, it is generally more difficult to find an antiderivative. In fact there are functions that do not have antiderivatives. However, the next three techniques allow us to find an antiderivative for many functions that are common in applications, but cannot be found from the basic formulas that we have listed above. The three techniques that we will discuss in the rest of this chapter on antiderivatives and indefinite integrals are change of variables, integration by parts, and partial fractions.

(#) <u>Integration using a Change of Variable</u>: This technique is often necessary in order to get an integral into a certain form that allows us to find the antiderivative. It is best to consider some examples.

<u>Example 1</u>: Find the $\int sin(3x^2)(x)\, dx$. This is best evaluated by letting

u = $3x^2$, so that du = (6x)dx. We have the "(x)dx" part already within the integrand, so that we just need to supply the 6 into the integrand and compensate by multiplying outside of the integral by $\frac{1}{6}$. So we have the result,

$\int sin(3x^2)(x)\, dx = \frac{1}{6}\int sin(3x^2)(6x)dx = \frac{1}{6}\int sin(u)du$,

which is in a form for which we have an integral formula that we can use. We have the integral formula:

$\int sin(x)du = (-cos(x) + C)$, so that

$\frac{1}{6}\int sin(u)du = \frac{1}{6}(-cos(u) + C) = \left(\frac{-1}{6}\right) cos(u) + C = \left(\frac{-1}{6}\right) cos(3x^2) + C$,

after going back to the variable x by substituting $(3x^2)$ for u.

Example 2: Find $\int \frac{e^{\sqrt{x}}}{\sqrt{x}}\, dx$. We let u = $\sqrt{x}$, and then du = $\frac{1}{2\sqrt{x}}\, dx$.

So we have $\int \frac{e^{\sqrt{x}}}{\sqrt{x}}\, dx = 2\int \frac{e^{\sqrt{x}}}{2\sqrt{x}}\, dx = 2\int e^u du$,

which we have an integral formula for:

We have the formula $\int e^x dx = e^x + C$.

So, $2\int e^u du = 2e^u + C = 2e^{\sqrt{x}} + C$,

after going back to the variable x.

Example 3: Find $\int sec^2(4x)\, dx$. Letting u = (4x), du = (4)dx, we have

$\frac{1}{4}\int sec^2(4x)(4)\, dx = \frac{1}{4}\int sec^2(u)\, du$,

which we have an integral formula for. We know:

$\int sec^2(x)\, dx = tan(x) + C$.

So, $\frac{1}{4}\int sec^2(u)\, du = \frac{1}{4}tan(u) + C = \frac{1}{4}tan(4x) + C$,

after going back to the variable x.

Example 4: Find $\int (4x^2 + 2x)^3 (4x + 1)\, dx$.

Let u = $4x^2 + 2x$, and then du = $(8x + 2)dx$. So, we have

$\frac{1}{2}\int (4x^2 + 2x)^3 (2)(4x + 1)\, dx = \frac{1}{2}\int (4x^2 + 2x)^3 (8x + 2)\, dx$

$= \frac{1}{2}\int u^3 du = \frac{1}{2}\left(\frac{u^4}{4}\right) + C = \frac{u^4}{8} + C$

$= \frac{(4x^2 + 2x)^4}{8} + C$, after going back to the variable x.

The preceding Examples 1 through 4 and Example 6 below, are for the most part, the most common type of change of variable problem that the student will encounter, in many applications and in the exercises following this section. In Example 5 we will go over a problem where we use a miscellaneous change of variable. This type of change of variable will not crop up at any later point in the book. Examples 7 and 8 below involve so-called trigonometric substitutions, which are very useful and important in many of the integrals that will occur in later exercises. The student should study these two examples well, and possibly you'll need to refer back to them to do several exercises later in the book. I advise you not to skip over exercises 13 and 14 at the end of this section.

Example 5: Find $\int \frac{x^2}{1+x}\, dx$. This is an integral that seems to be

quite unusual, that is, it isn't even close to a form that we know. However, a simple change of variable will work wonders.
Let $u = 1 + x$, then $x = u - 1$, $x^2 = (u^2 - 2u + 1)$, and dx = du.

Then $\int \frac{x^2}{1+x}\, dx = \int \frac{u^2 - 2u + 1}{u}\, du = \int (u - 2 + \frac{1}{u})\, du$

$= \frac{u^2}{2} - 2u + ln|u| + C = \frac{(x+1)^2}{2} - (2x+2) + ln|x+1| + C$,
after going back to the variable x.

Example 6: Find $\int \frac{ln(x)}{x}\, dx$. Let $u = ln(x)$, and then $du = \frac{1}{x}\, dx$.

Then, $\int \frac{ln(x)}{x}\, dx = \int (u)\, du = \frac{u^2}{2} + C = \frac{(lnx)^2}{2} + C$,

after going back to the variable x.

Example 7: Find $\int \frac{1}{\sqrt{1+x^2}}\, dx$. This integral seems quite unusual at first

glance, but with a simple change of variable technique things will work out. Remembering that $1 + tan^2(x) = sec^2(x)$ and $sin^2(x) + cos^2(x) = 1$ is often very useful. The technique that we will use here is called a trigonometric substitution.

Let x = tan(θ), dx = $sec^2(\theta)\, d\theta$, $\theta = tan^{-1}(x)$.

Then $\int \frac{1}{\sqrt{1+x^2}}\, dx = \int \frac{1}{\sqrt{1+tan^2\theta}} sec^2\theta\, d\theta$.

Making use of the trigonometric identity $1 + tan^2\theta = sec^2\theta$, the integral

$\int \frac{1}{\sqrt{1+x^2}}\, dx = \int \frac{1}{sec\theta} sec^2\theta\, d\theta = \int sec(\theta)\, d\theta = ln|sec\theta + tan\theta| + C$

= ln|sec $(tan^{-1}(x)) + tan(tan^{-1}(x))| + C$, after going back to the variable x.

Now, refer to the illustration below:

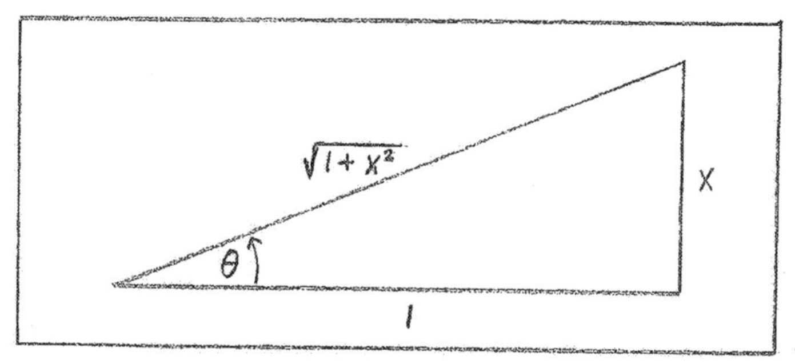

The $tan^{-1}(x)$ is the angle θ such that tan(θ) = x.

From the above right triangle, we can see that the hypotenuse is $\sqrt{1+x^2}$.

So, the sec(θ) $= \sqrt{1+x^2}$, and the integral $\int \frac{1}{\sqrt{1+x^2}}\, dx$ = ln| $\sqrt{1+x^2} + x| + C$.

<u>Example 8</u>: Find $\int \sqrt{1-x^2}\, dx$.

Once again, a couple changes of variable are in order.

Let x = sin(θ), dx = cos $(\theta)\, d\theta$. Then making use of the trigonometric identity $sin^2\theta + cos^2\theta = 1$,

$\int \sqrt{1-x^2}\, dx = \int \sqrt{cos^2\theta} \cdot cos\theta\, d\theta = \int cos^2\theta\, d\theta$.

Now use the trigonometric identity $cos^2\theta = \frac{1+cos(2\theta)}{2}$. So we have

$\int cos^2\theta\, d\theta = \int \frac{1+cos(2\theta)}{2}\, d\theta = \frac{1}{2}\int d\theta + \frac{1}{2}\int cos(2\theta)\, d\theta$.

Now let u = 2θ, du = $2d\theta$ in the second integral to get ,

$\frac{1}{2}\int d\theta + \frac{1}{2}\int cos(2\theta)\, d\theta \;=\; \frac{1}{2}\theta + \frac{1}{4}\int cos(2\theta)(2)\, d\theta = \frac{1}{2}\theta + \frac{1}{4}\int cos(u)\, du$

$= \frac{1}{2}\theta + \frac{1}{4}sin(u) + C = \frac{1}{2}\theta + \frac{1}{4}sin(2\theta) + C$

$= \frac{1}{2}\theta + \frac{1}{4}(2sin\theta cos\theta) + C$, where I have used the trigonometric identity $sin(2\theta) = 2\,sin\theta cos\theta$.

Now going back to the variable x, where $\theta = sin^{-1}(x)$, we have

$\frac{1}{2}\theta + \frac{1}{4}(2sin\theta cos\theta) + C = \frac{1}{2}sin^{-1}(x) + \frac{1}{2}sin(sin^{-1}(x))cos(sin^{-1}(x)) + C$

Referring to the illustration below, we see that the last expression simplifies to

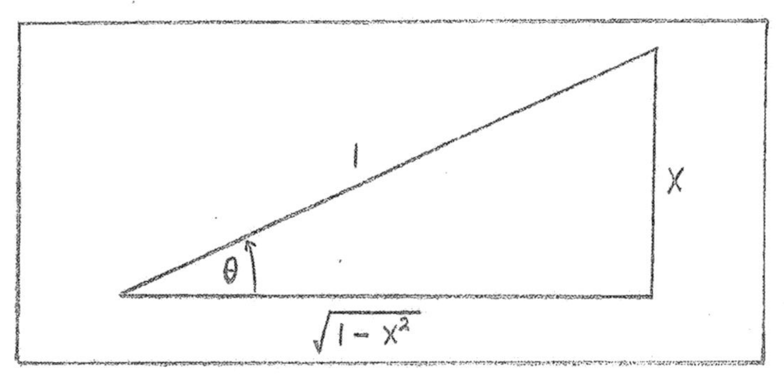

$\frac{1}{2}(sin^{-1}(x) + (x)(\sqrt{1-x^2})) + C$

The reader should realize from the above examples that change of variable techniques can require a bit of ingenuity, and a lot of practice.

EXERCISES:

Evaluate the following integrals:

(1.) $\int sin(3x^2)(x)\, dx$

(2.) $\int (4x^3 - 9)^5 (x^2)\, dx$

(3.) $\int sec(5x)\,dx$

(4.) $\int \frac{1}{7x}\,dx$

(5.) $\int e^{13x}\,dx$

(6.) $\int e^{(13x^2+1)}(x)\,dx$

(7.) $\int \frac{x}{7x^2-12}\,dx$

(8.) $\int (8x^2-16x)^3(x-1)\,dx$

(9.) $\int (\sqrt{x}+2)^5 \cdot \frac{1}{\sqrt{x}}\,dx$

(10.) $\int (10x^2-20x)^2(x-1)\,dx$

(11.) $\int tan(5x)\,dx$

(12.) $\int cos(x^2+2)(x)\,dx$

(13.) $\int \frac{1}{\sqrt{1-16x^2}}\,dx$

(14.) $\int \frac{1}{1+25x^2}\,dx$

(15.) $\int 3x\sqrt{1+x^2}\,dx$

(16.) $\int (37)\sqrt{1+3x^2}\,(x)\,dx$

(17.) $\int (x)\,e^{x^2}\,dx$

(18.) $\int \frac{1}{8x+2}\,dx$

(19.) $\int \frac{1}{(3x+2)^3}\,dx$

(20.) $\int \sqrt[4]{2x^2+1}\,(x)\,dx$

(#) <u>Integration by Parts</u>: This is a very important and commonly used integration technique. It is derived from the product rule for differentiation.

The product rule for differentiating the product of two functions u(x) and v(x), as you should recall, is

$$\frac{d}{dx}(u(x)\cdot v(x)) = u(x)\cdot v'(x) + v(x)\cdot u'(x)\ .$$

Multiplying through by dx,

$$d(u(x)\cdot v(x)) = u(x)\cdot v'(x)dx + v(x)\cdot u'(x)dx\ .$$

Integrating both sides, we have,

$$\int d(u(x)\cdot v(x)) = \int u(x)\cdot v'(x)dx + \int v(x)\cdot u'(x)dx$$

Denoting u(x) and v(x) simply as u and v respectively, we have

$\int d(uv) = \int uv'dx + \int vu'dx$ or $(uv) = \int udv + \int vdu$

From this we have the integration by parts formula:

$$\int udv = uv - \int vdu \ .$$

To use this formula we have to identify the part of the integrand that we will call the function u and the remaining part of the integrand (including the differential) that we will call dv. Then we find du (from u) and v (by integrating dv). Then we put all of the parts into the formula and evaluate it.

<u>Example 1</u>: Find $\int xe^{-x}\, dx$.

Let u = x, dv = $e^{-x}dx$

Then, du = dx, v = $\int dv = \int e^{-x}dx = -e^{-x}$

(Note: we do not add an arbitrary constant C here in figuring "v")

So, $\int xe^{-x}\, dx$

$= (-xe^{-x}) + \int e^{-x}dx$

$= (-xe^{-x} - e^{-x} + C) = (-e^{-x})(x+1) + C$.

<u>Example 2</u>: Find $\int x^2(ln(x))\, dx$.

Let u = ln(x), dv = x^2dx

Then, du = $\frac{1}{x}dx$, v = $\int dv = \int x^2\, dx = \frac{x^3}{3}$.

So, $\int x^2(ln(x))\,dx$

$= (\frac{1}{3}x^3 lnx) - \frac{1}{3}\int x^2\,dx$

$= (\frac{1}{3}x^3 lnx) - \frac{1}{9}x^3 + C \;=\; (\frac{1}{3}x^3)(lnx - \frac{1}{3}) + C$.

Example 3: Find $\int (x)(tan^{-1}(x))\,dx$.

Let u = $tan^{-1}(x)$, dv = x dx
Then, du = $\frac{1}{1+x^2}\,dx$, v = $\frac{x^2}{2}$.

So, $\int (x)(tan^{-1}(x))\,dx \;=\; (\frac{x^2}{2}\,tan^{-1}(x)) - \frac{1}{2}\int \frac{x^2}{1+x^2}\,dx$

$= (\frac{x^2}{2}\,tan^{-1}(x)) - \frac{1}{2}\int (1 - \frac{1}{1+x^2})\,dx$ (after using long division)

$= \frac{1}{2}x^2\,tan^{-1}(x) - \frac{1}{2}x + \frac{1}{2}\,tan^{-1}(x) + C$
$= \frac{1}{2}\,tan^{-1}(x)\,(x^2+1) - \frac{1}{2}x + C$.

Example 4: Find $\int x^n sin(x)\,dx$, n is a positive integer.

Let u = x^n , dv = sin(x) dx ,

Then du = $nx^{n-1}\,dx$, v = $\int sin(x)\,dx = -cos(x)$.

So, $\int x^n sin(x)\,dx \;=\; -x^n cos(x) + n\int x^{n-1} cos(x)\,dx + C$.

Therefore, the reader can see that n iterations of this technique will give you the required result.

(#) Partial Fraction Decomposition: This technique is useful when we have a rational function $\frac{P(x)}{Q(x)}$ (a ratio of two polynomials with rational coefficients), where Q(x) can be factored into a product of linear and irreducible quadratic factors. An irreducible quadratic factor is one of the form $(ax^2 + bx + c)$, where $(b^2 - 4ac) < 0$.

Firstly, the degree of P(x) must be less than the degree of Q(x). If this is not the case, then one must do long division until it is so. Then, if Q(x) can be factored into r distinct linear factors of multiplicities $m_1, m_2, \ldots, m_r$ and s distinct irreducible quadratic factors of multiplicities $n_1, n_2, \ldots, n_s$, then there are theorems in the field of algebra that say,

$\frac{P(x)}{Q(x)}$ = a sum of $(m_1 + \cdot\cdot\cdot + m_r + n_1 + \cdot\cdot\cdot\ n_s)$ partial fractions.
Let's describe what these partial fractions are.

For every linear factor of the form $(x - a)^m$, there are m fractions of the form $\frac{A_1}{(x-a)} + \cdot\cdot\cdot + \frac{A_m}{(x-a)^m}$.

For every irreducible quadratic factor of the form $(ax^2 + bx + c)^n$, there are n fractions of the form
$\frac{B_1 + C_1 x}{(ax^2 + bx + c)} + \cdot\cdot\cdot + \frac{B_n + C_n x}{(ax^2 + bx + c)^n}$.

All of the constants A_i, B_i, C_i, and so forth, need to be determined, and then the $\int \frac{P(x)}{Q(x)} dx = \int$(the sum of the partial fractions) dx. Usually, the antiderivatives of all the partial fractions are easily found. It is finding the constants that may require the most effort, depending on the complexity of the problem.

It is best to consider some examples:

Example 1: Find the $\int \frac{2x+4}{x^2+6x}\, dx$. This can be written $\int \frac{2x+4}{x(x+6)}\, dx$.

The first task is to find the partial fraction decomposition for the integrand: $\frac{2x+4}{x(x+6)} = \frac{A}{x} + \frac{B}{(x+6)}$. To find A and B, multiply both sides by x(x + 6) to get

2x + 4 = A(x + 6) + B(x) = (A + B)x + (6A)

Equating coefficients, we have (A + B) = 2 and (6A) = 4.

Therefore, A = $\frac{2}{3}$ and B = 2 - $\frac{2}{3}$ = $\frac{4}{3}$, so that $\frac{2x+4}{x(x+6)} = \frac{(\frac{2}{3})}{x} + \frac{(\frac{4}{3})}{(x+6)}$

So, $\int \frac{2x+4}{x^2+6x}\,dx = \int \frac{2x+4}{x(x+6)}\,dx$

$= \frac{2}{3}\int \frac{1}{x}\,dx + \frac{4}{3}\int \frac{1}{(x+6)}\,dx$

$= \frac{2}{3}\,ln|x| + \frac{4}{3}\,ln|x+6| + C$.

Example 2: Find $\int \frac{3x^2-2x+1}{x^4+6x^3+10x^2+6x+9}\,dx$.

It turns out that $(x^4+6x^3+10x^2+6x+9) = (x+3)^2(x^2+1)$. So the denominator of the integrand has been factored into linear and irreducible quadratic factors.

So, we want to evaluate $\int \frac{3x^2-2x+1}{(x+3)^2(x^2+1)}\,dx$. The partial fraction decomposition of the integrand is $\frac{3x^2-2x+1}{(x+3)^2(x^2+1)} = \frac{A}{(x+3)} + \frac{B}{(x+3)^2} + \frac{Cx+D}{(x^2+1)}$. We need to find the constants A, B, C, and D.

Now, multiplying through by $(x+3)^2(x^2+1)$ gives us:

$3x^2-2x+1 = A(x+3)(x^2+1) + B(x^2+1) + (Cx+D)(x+3)^2$

$= A(x^3+3x^2+x+3) + Bx^2 + B + (Cx+D)(x^2+6x+9)$

$= Ax^3 + 3Ax^2 + Ax + 3A + Bx^2 + B + Cx^3 + Dx^2 + 6Cx^2 + 6Dx + 9Cx + 9D$

$= (A+C)x^3 + (3A+B+6C+D)x^2 + (A+9C+6D)x + (3A+B+9D)$

After equating coefficients, we have,

$A+C=0$

$3A+B+6C+D=3$

$A+9C+6D=-2$

$3A+B+9D=1$

I want to eliminate A and B from the second equation:

Subtracting the fourth equation from the second, we have $6C-8D=2$.

So then $\frac{8}{6}(6C-8D=2) \Rightarrow 8C-(\frac{32}{3})D=\frac{8}{3}$.

I want to to eliminate A from the third equation:

Letting $A=-C$ in the third equation leads to $8C+6D=-2$.

So that we have: $8C-(\frac{32}{3})D=\frac{8}{3}$ and $8C+6D=-2$, involving C and D. Subtracting the second of these two from the first eliminates C and enables us to determine D. Then we will be able to determine C:

So we have $(\frac{-32}{3}-\frac{18}{3})D=\frac{14}{3} \Rightarrow -50D=14 \Rightarrow D=\frac{-7}{25}$.

Therefore, $8C=(\frac{-50}{25}+\frac{42}{25})=\frac{-8}{25} \Rightarrow C=\frac{-1}{25}$.

So now, $A=-C=\frac{1}{25}$, from the first of the original equations.

And finally, $B=1-3A-9D=\frac{25}{25}-\frac{3}{25}+\frac{63}{25}=\frac{85}{25}=\frac{17}{5}$.

In summary, $A=\frac{1}{25}$, $B=\frac{17}{5}$, $C=\frac{-1}{25}$, $D=\frac{-7}{25}$. A piece of cake!

So, $\int\frac{3x^2-2x+1}{x^4+6x^3+10x^2+6x+9}\,dx = \int\frac{3x^2-2x+1}{(x+3)^2(x^2+1)}\,dx$

$$= \frac{1}{25}\int\frac{1}{x+3}\,dx + \frac{17}{5}\int(x+3)^{-2}dx + \int\frac{\frac{-1}{25}x-\frac{7}{25}}{x^2+1}\,dx$$

$$= \frac{1}{25}ln|x+3| - \frac{17}{5}\left(\frac{1}{x+3}\right) - \frac{1}{25}\left(\frac{1}{2}\int\frac{2x}{x^2+1}dx + 7\int\frac{1}{x^2+1}dx\right)$$

$= \frac{1}{25} ln|x+3| - \frac{17}{5}\left(\frac{1}{x+3}\right) - \frac{1}{25}\left(\frac{1}{2} ln|x^2+1| + 7tan^{-1}(x)\right)$
$= \frac{1}{25} ln|x+3| - \frac{17}{5}\left(\frac{1}{x+3}\right) - \frac{1}{50} ln|x^2+1| - \frac{7}{25} tan^{-1}(x) + C.$

Example 3: Find $\int \frac{5x^2 - x - 1}{5x^3 + 10x^2 + 5x}\, dx$.

The denominator can be factored $5(x)(x^2 + 2x + 1) = 5(x)(x+1)^2$.

So, $\int \frac{5x^2 - x - 1}{5x^3 + 10x^2 + 5x}\, dx = \frac{1}{5}\int \frac{5x^2 - x - 1}{(x)(x+1)^2}\, dx$.

The partial fraction decomposition of the integrand is,
$\frac{5x^2 - x - 1}{(x)(x+1)^2} = \frac{A}{x} + \frac{B}{(x+1)} + \frac{C}{(x+1)^2}$. We need to determine A, B, and C.

Multiply through by $(x)(x+1)^2$. Then we have,

$$5x^2 - x - 1 = A(x+1)^2 + B(x)(x+1) + C(x)$$
$$= Ax^2 + 2Ax + A + Bx^2 + Bx + Cx$$
$$= (A+B)x^2 + (2A+B+C)x + A$$

Equating coefficients: $A + B = 5$

$2A + B + C = -1$

$A = -1$

So $A = -1$, $B = 6$,
$C = (-1 - 2A - B) = (-1 + 2 - 6) = -5$.

So $\frac{5x^2 - x - 1}{(x)(x+1)^2} = \frac{-1}{x} + \frac{6}{(x+1)} + \frac{-5}{(x+1)^2}$.

Then $\int \frac{5x^2 - x - 1}{5x^3 + 10x^2 + 5x}\, dx = \frac{1}{5}\int \frac{5x^2 - x - 1}{(x)(x+1)^2}\, dx$

$$= \frac{1}{5}\left(\int \frac{-1}{x}\, dx + \int \frac{6}{(x+1)}\, dx + (-5)\int (x+1)^{-2} dx\right)$$
$$= \frac{-1}{5} ln|x| + \frac{6}{5} ln|x+1| + \frac{1}{(x+1)} + C.$$

EXERCISES:

Evaluate the following integrals:

(1.) $\int xe^x \, dx$

(2.) $\int 3x \cos(x) \, dx$

(3.) $\int \sqrt{x} \ln(x) \, dx$

(4.) $\int x^2 e^x \, dx$

(5.) $\int (4x) \sin(4x) \, dx$

(6.) $\int \ln(x) \, dx$

(7.) $\int \frac{x+1}{x^2 - 9x} \, dx$

(8.) $\int \frac{1}{(x)(x+1)} \, dx$

(9.) $\int \frac{30x}{(x-1)(x^2+1)} \, dx$

(10.) $\int \frac{15}{(x^2+1)} \, dx$

(11.) $\int \frac{3x^3 + 2x + 1}{x^2 + 2x + 1} \, dx$

(12.) $\int \frac{4x^2 - x - 12}{x^2 + x} \, dx$

(9) The Definite Integral

(#) **Definite Integrals**: Up to this point we have been dealing with indefinite integrals, which we know to be antiderivatives. We will now learn how to use the antiderivative to solve certain problems that are important and useful to mathematicians and scientists. We do this by considering what we call a definite integral. The expression

$$\int_a^b y(x)\, dx$$

is called a definite integral with lower limit "a" and upper limit "b." We can assume that y(x) $\geq$ 0 on the closed interval [a,b]. Then $\int_a^b y(x)\, dx$ is the area under the curve y(x), above the x-axis, and between the two vertical lines x = a and x = b, as shown in the figure:

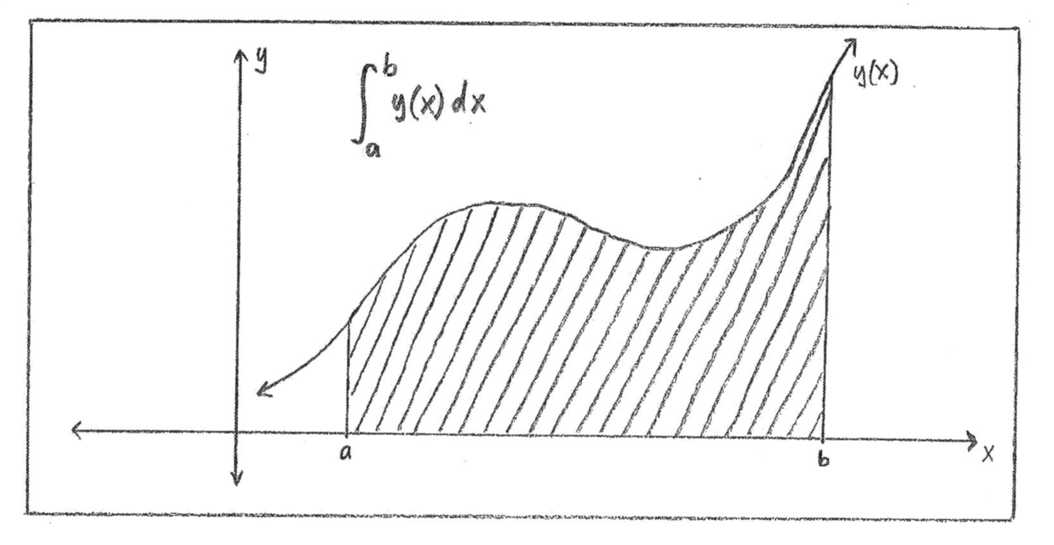

Definite integrals can be used to find areas, lengths of curves, volumes, and other things of a geometric nature, as we shall see. The integral often has a meaning other than just the geometric meaning. Definite integrals are very important in the application of calculus to the sciences and engineering.

(#) **Riemann Sums**: For $\int_a^b y(x)\,dx$, on the closed interval [a,b],

Let $\Delta x = (\frac{b-a}{n})$. This breaks the interval [a,b] into a partition of n equal segments, where the boundaries of the segments are the set of (n + 1) points $a = x_0 < x_1 < x_2 < \cdot\cdot\cdot < x_{n-2} < x_{n-1} < x_n = b$. Then if the point c_i is any point in $[x_{i-1}, x_i]$, we have n rectangles of height y(c_i) and width $\Delta x = (x_i - x_{i-1}) = (\frac{b-a}{n})$. The sum of the areas of these n rectangles is an approximation to the exact area under the curve, which is given by the integral $\int_a^b y(x)\,dx$. So we have $\int_a^b y(x)\,dx = \lim_{n\to\infty} \sum_{i=1}^{n} y(c_i)\Delta x$, if the limit exists. This type of integral is called a Riemann integral, named after the 19th century German mathematician Bernhard Riemann, and the summation is called a Riemann sum. See the figure below. A function such that this integral exists is said to be Riemann integrable on [a,b]. This way of formulating an integral is sufficient for just about any application that arises. This definite integral will always exist if y(x) is continuous on a closed interval [a,b]. It may or may not exist on an open interval (a,b), depending on the nature of y(x) on this interval, such as if y(x) has vertical asymptotes at x = a or x = b, or at any number between a and b.

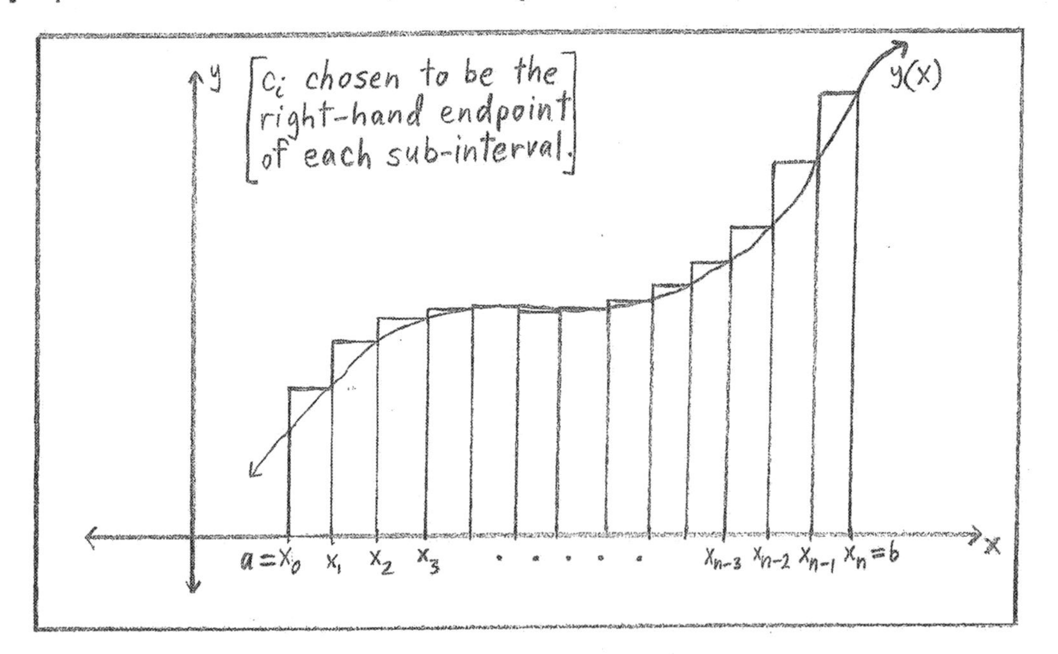

Here are a few important properties of definite integrals which are very useful in proving the Fundamental Theorem of Calculus in the next section.

(A) $\int_a^b y(x)\,dx = \lim_{n\to\infty} \sum_{i=1}^{n} y(c_i)(x_i - x_{i-1})$ is the formulation of the integral on [a,b] with the partition as described above with a < b. If we have the partition reversed where $b = x_0 > x_1 > x_2 > \cdot \cdot \cdot > x_{n-1} > x_n = a$, then we would have $\int_b^a y(x)\,dx = \lim_{n\to\infty} \sum_{i=1}^{n} y(c_i)(x_i - x_{i-1}) = -\int_a^b y(x)\,dx$, because $(x_i - x_{i-1})$ for the second partition would be the negative of $(x_i - x_{i-1})$ in the usual partition. The $y(c_i)$ would be the same in both cases. However, with the usual formulation, $x_{i-1} \le c_i \le x_i$, but with things turned around $x_i \le c_i \le x_{i-1}$. So summarizing, we have $\int_a^b y(x)\,dx = -\int_b^a y(x)\,dx$.

(B) $\int_a^b y(x)\,dx = \int_a^c y(x)\,dx + \int_c^b y(x)\,dx$, where c is any constant such that the function y(x) is continuous on some interval containing a,b, and c.

(C) $\int_a^a y(x)\,dx = 0$, for every number a.

(#) <u>The Fundamental Theorem of Calculus</u>:

Let y(x) be continuous on the closed interval [a, b]. Let x be any point in (a,b). Then Y(x) = $\int_a^x y(t)\,dt$ is the area shown in the figure below,

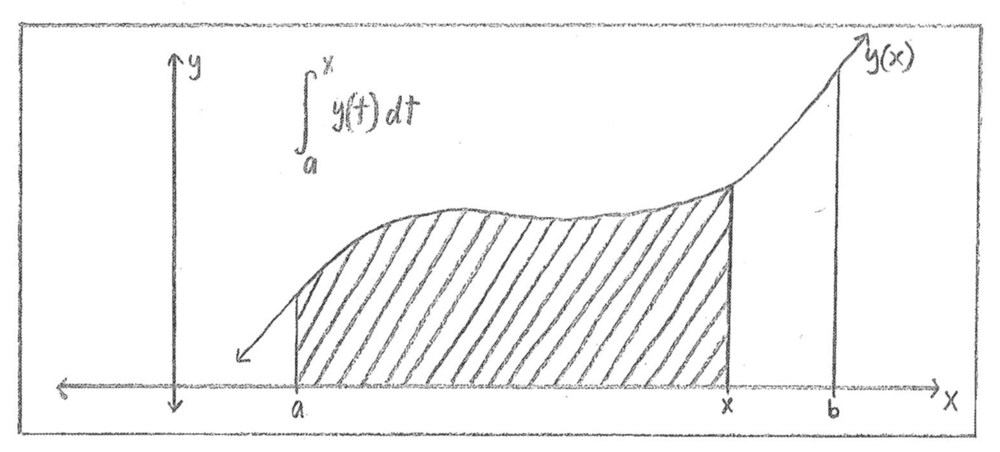

In this expression for Y(x), the variable x is the upper limit. The variable t is called a dummy variable, since we could just as well use any letter in place of it. The Fundamental Theorem of Calculus says that:

(A) Y(x) is an antiderivative of y(x).

(B) $\int_a^b y(x)\,dx = Y(b) - Y(a)$, for any antiderivative Y(x) of y(x).

Proof:

(A) If Y(x) = $\int_a^x y(t)\,dt$, then

$$Y'(x) = \lim_{h\to 0} \frac{Y(x+h) - Y(x)}{h} = \lim_{h\to 0} \frac{\int_a^{x+h} y(t)\,dt - \int_a^x y(t)\,dt}{h}$$

$$= \lim_{h\to 0} \frac{\int_x^a y(t)\,dt + \int_a^{x+h} y(t)\,dt}{h} = \lim_{h\to 0} \frac{\int_x^{x+h} y(t)\,dt}{h} .$$

Since y(x) is continuous on [x, x+h], then clearly there is some number M such that $\int_x^{x+h} y(t)\,dt = (M \cdot h)$, as shown in the figure below, where the area of the rectangle defined by vertical lines at x and x+h, and the horizontal lines y = 0 and y = M, is equal to the area under the curve y(x) from x to x+h, denoted by $\int_{x}^{x+h} y(t)\,dt$. M is called the mean value.

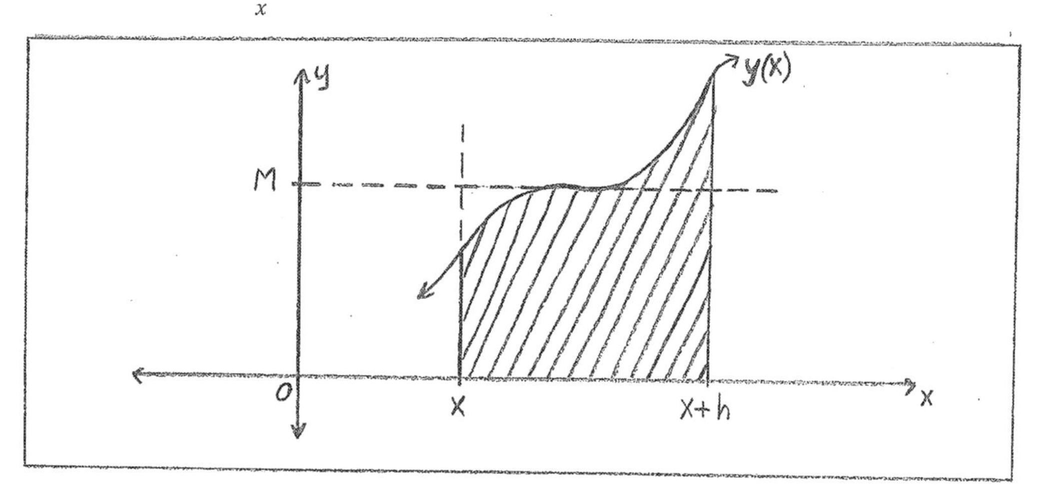

Then Y'(x) = $\lim_{h\to 0}(\frac{M\cdot h}{h}) = \lim_{h\to 0}(M) = y(x)$. This is clear since the mean value M $\to y(x)$ as $h \to 0$.

Therefore, Y(x) is an antiderivative of y(x).

(B) Let c be any element of [a,b], and let Y(x) = $\int_{c}^{x} y(t)\,dt$ be a function of x for all x in [a,b]. We know from part (A) that Y(x) is an antiderivative of y(x). The starting point, the lower limit, is not important to this result. What is important is that the rate of change of the "accumulation of area," Y'(x), at the right-hand endpoint x, is just y(x). Many important problems in science and engineering which have formulations defined by definite integrals are easily solved because all we need to do is find an antiderivative of y(x), and then the task of evaluating the definite integral is easy:

The $\int_{a}^{b} y(x)\,dx = \int_{a}^{c} y(x)\,dx + \int_{c}^{b} y(x)\,dx$

$= \int_{c}^{b} y(x)\,dx - \int_{c}^{a} y(x)\,dx$

$= Y(b) - Y(a)$.

So to evaluate $\int_a^b y(x)\,dx$; we find an antiderivative Y(x) for y(x) and then evaluate it between the limits a and b, that is, calculate the difference $(Y(b)-Y(a))$. Geometrically, this is the area under y(x) and above the x-axis from x = a to x = b. If F(x) and G(x) are any two antiderivatives of y(x), then they differ at most by a constant C, which means that G(x) = F(x) + C. Then clearly the difference $G(b)-G(a) = (F(b)+C) - (F(a)+C) = (F(b)-F(a))$. We can see that the constant C simply cancels out.

This proves the Theorem.

EXERCISES:

Evaluate the following definite integrals.

(1.) $\int_0^1 x^2\,dx$

(2.) $\int_2^4 \sqrt{x}\,dx$

(3.) $\int_0^{\frac{\pi}{2}} sinx\,dx$

(4.) $\int_{\pi}^{\frac{3\pi}{2}} cosx\,dx$

(5.) $\int_1^3 \frac{x^3-x}{2}\,dx$

(6.) $\int_0^1 \frac{1}{\sqrt{1-x^2}}\,dx$

(7.) $\int_{\frac{\pi}{6}}^{\frac{\pi}{4}} sec\theta\,d\theta$

(8.) $\int_6^{13} \sqrt{x}+\frac{1}{2}x\,dx$

(9.) $\int_{-1}^1 |x|\,dx$

(10.) $\int_{-2}^0 cosx\,dx$

(11.) $\int_2^5 2x^2+3x^3$ dx

(12.) $\int_1^4 \frac{\sqrt{x}-x}{\sqrt{x}}\,dx$

(10) Applications of Indefinite and Definite Integrals

(#) Areas Between Curves: There are many applications of integration. We will start by considering the geometric problem of calculating the area between two curves over some range of x-values. The area between two curves (many times the lower curve is the x-axis) can have some meaning other than just its geometric meaning. However, in the four examples of this section, we will just consider the geometric problem of finding the area of a region defined by some curves.

Example 1: Find the area between y = x^3 and $y = -x$, from x = 0 to 1. This area is shaded in the figure below,

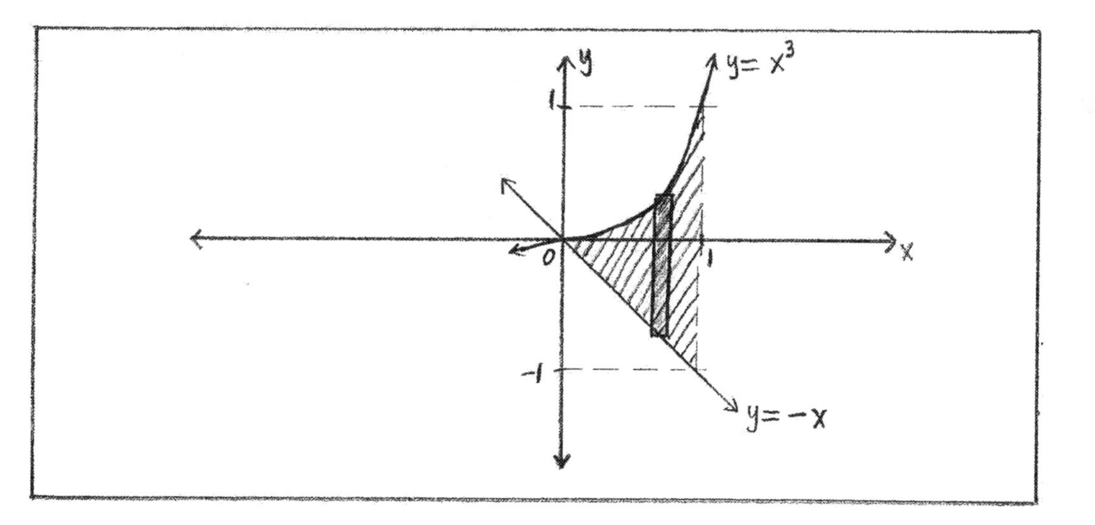

The area between two curves is conceptualized as a Riemann sum where the number of rectangles goes to infinity, and the rectangles extend from the lower function to the higher function on a certain range of x-values. In

the Riemann sum the height of each rectangle is then the upper function evaluated at a point $(c_i) \in [x_{i-1}, x_i]$ minus the lower function evaluated at the same point (c_i).

So the area = $\int_0^1 (x^3 - (-x))\, dx = \int_0^1 (x^3 + x)\, dx$

$= \left[\frac{x^4}{4} + \frac{x^2}{2}\right]_0^1 = \frac{1}{4} + \frac{1}{2} = \frac{3}{4}$.

Example 2: Find the area shaded in the figure below, which involves the sine and cosine functions. The shaded area involves the area under y = sin(x) and above the x-axis, from 0 to $\frac{\pi}{4}$, and under y = cos(x) and above the x-axis, from $\frac{\pi}{4}$ to $\frac{\pi}{2}$. These two areas must be considered separately because we have two different functions defining the upper curve.

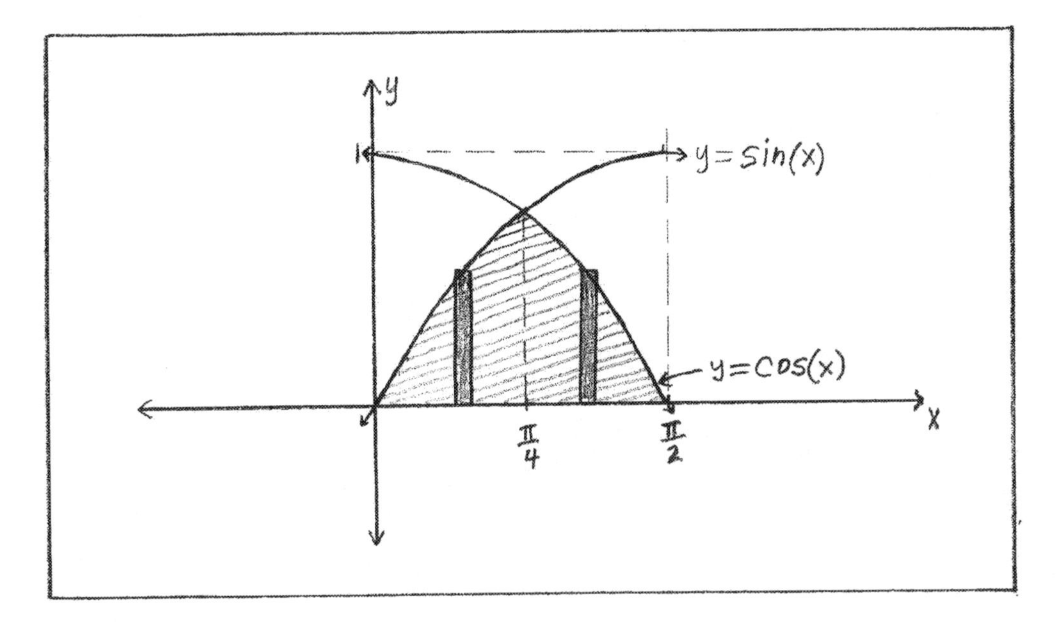

The Area = $\int_0^{\frac{\pi}{4}} sin(x)\, dx + \int_{\frac{\pi}{4}}^{\frac{\pi}{2}} cos(x)\, dx = [-cos(x)]_0^{\frac{\pi}{4}} + [sin(x)]_{\frac{\pi}{4}}^{\frac{\pi}{2}}$

$= (-cos(\frac{\pi}{4}) + cos(0)) + (sin(\frac{\pi}{2}) - sin(\frac{\pi}{4}))$

$= (1 - \frac{1}{\sqrt{2}}) + (1 - \frac{1}{\sqrt{2}}) = (2 - \sqrt{2}) \approx 0.586$

Example 3: Find the area between y = $\frac{1}{10}x^2$ and $y = 3$ from $-\sqrt{30}$ to $\sqrt{30}$. These are the x-values which correspond to the points of intersection of these curves. This area is shaded in the figure below.

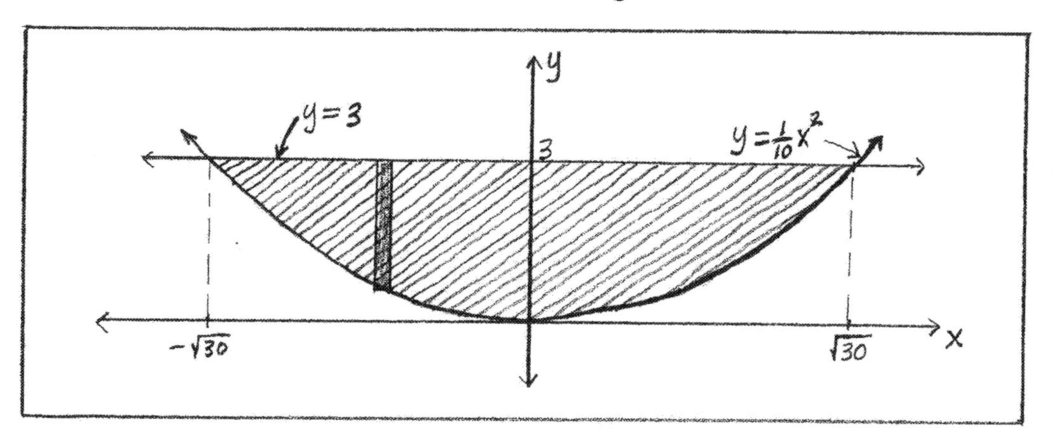

The area is $A = \int_{-\sqrt{30}}^{\sqrt{30}} (3 - \frac{1}{10}x^2)\,dx = [3x - \frac{1}{30}x^3]_{-\sqrt{30}}^{\sqrt{30}}$

$= ((3\sqrt{30}) - \frac{1}{30}(30\sqrt{30})) - ((-3\sqrt{30}) + \frac{1}{30}(30\sqrt{30}))$

$= (4\sqrt{30}) \approx 21.91$

Example 4: Find the area between $y = 2 - x$ and $y = x - 3$, and to the right of the y-axis. This area is shaded in the figure below.

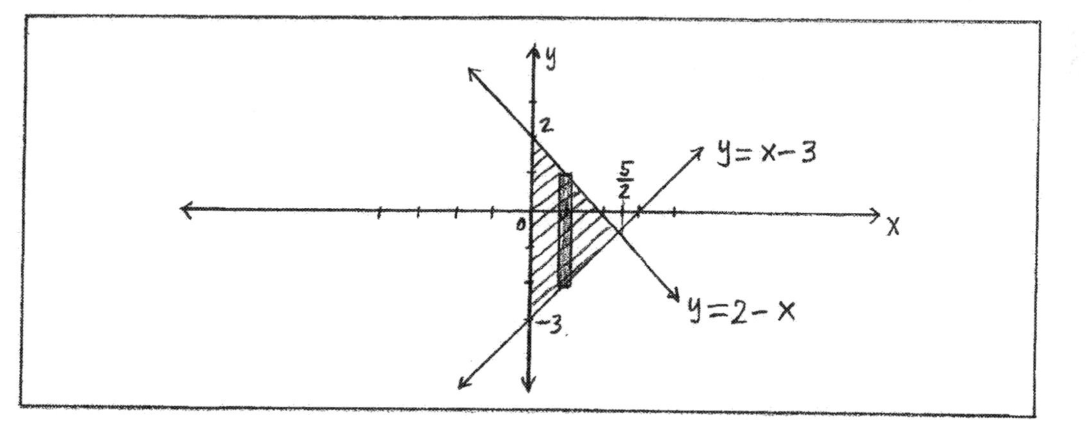

To find the x-value corresponding to the point of intersection of these two lines, solve $2 - x = x - 3$ to get $x = \frac{5}{2}$. Then the area is determined by solving the integral:

$$\int_0^{\frac{5}{2}}[(2-x)-(x-3)]\,dx = \int_0^{\frac{5}{2}}(5-2x)\,dx = [5x-x^2]_0^{\frac{5}{2}} = (\tfrac{25}{2}-\tfrac{25}{4}) = \tfrac{25}{4} = 6.25$$

EXERCISES:

(1.) What is the area under $y = x^3$, above the x-axis, and between x = 1 and x = 3?

(2.) Find the area between $y = \frac{1}{x}$ and $y = \frac{1}{x^2}$ from x = 1 to x = 6.

(3.) Find the area under the curve $y = e^{-2x}$ and above the x-axis from x = 0 to x = 10.

(4.) What is the area under $y = tanx$ from x = 0 to x = $\frac{\pi}{3}$?

(5.) What is the area between y = sinx and y = -sinx from x = 0 to x = 5?

(6.) Find the area between y = 2x and y = x + 10 and the y-axis.

(7.) Find the area between y = x and y = 2x from x = -5 to x = 0.

(#) Some Applications to Physics:

(Position, Velocity, Acceleration)

Classical and modern physics would not exist without calculus, and therefore we would not have our modern civilization without it. We would still be living essentially as we did many years ago, without all of our modern conveniences. Fortunately, the world is more comprehensible than was imagined, and advanced math was an inevitable development of the European Renaissance.

The first application to physics that we will consider is just simple non-uniform motion in one dimension. In one dimension, we can model this as motion restricted to the real number line, the x-axis or the y-axis. In the analysis of motion discovered by physicists in the 1600's, there evolved three main mathematical concepts. These three mathematical objects we call a position function of time x(t), a velocity function of time v(t), and an acceleration function of time a(t).

The position function gives us the point on the x-axis where some object is at time t. From this we can define the average change in position between times t = a and t = b, or average velocity on this time interval, as

$$v_{av} = \frac{x(b) - x(a)}{b - a} = \frac{\Delta x}{\Delta t}.$$

If we let "a" be fixed and have "b" approach a, then the average velocity v_{av} converges to the derivative $\frac{dx}{dt}$ at time t = a. So the velocity function

$$v(t) = \frac{dx}{dt}.$$

Now consider the average change in velocity between times t = a and t = b, or average acceleration on this time interval,

$$a_{av} = \frac{v(b) - v(a)}{b - a} = \frac{\Delta v}{\Delta t}.$$

Since "a" is fixed and "b" approaches a, then the average acceleration a_{av} converges to the derivative $\frac{dv}{dt}$ at time t = a. So the acceleration function

$$a(t) = \frac{dv}{dt}.$$

We have started with a position function and by two differentiations arrived at the acceleration function.

Now we can go in the opposite direction and start with the acceleration function a(t). For simplicity's sake, assume that the acceleration function is a constant function a(t) = g. Since $a(t) = \frac{dv}{dt}$, then $dv = a(t)\,dt$. If we integrate both sides, $\int dv = \int a(t)\,dt$. This leads to $v = \int g\,dt = gt + C$.

Usually we have an initial condition at time t = 0, v(0) = some constant v_0, which then means that C = v_0. So that finally we have the velocity function,

$$v(t) = gt + v_0.$$

Now since, $v(t) = \frac{dx}{dt}$, then $dx = v(t)\,dt$. If we integrate both sides, $x = \int dx = \int v(t)\,dt = \int (gt + v_0)\,dt = \frac{gt^2}{2} + v_0 t + K$. Usually we also have an initial condition at time t = 0 for position, which says x(0) = some constant x_0, which then means that K = x_0. So then we have the position function,

$$x(t) = \frac{gt^2}{2} + v_0 t + x_0.$$

Obviously, things could be more complicated if we had started with a more complicated acceleration function. Usually, in the kinematics of a moving object often studied in elementary physics , we make the assumption that the acceleration of gravity near the Earth's surface is essentially constant, which leads to an analysis of projectile motion near the Earth's surface along the lines of what we have just done. Since gravity is the only force acting on a projectile near the Earth's surface (ignoring air resistance), the above analysis is how motion in the vertical direction is modeled.

Example 1: Assume that a mass moves along the x-axis with an acceleration a(t) = 5t, for t $\geq$ 0. Let v(0) = 100 ft./sec., and x(0) = 50 ft. be our initial conditions. Find the velocity and position functions.

The velocity function is v(t) = $\int a(t)\,dt = \int (5t)\,dt = \frac{5}{2}t^2 + C$. From the initial condition v(0) = 100, we have v(0) = $\frac{5}{2}(0)^2 + C = 100$. Therefore C = 100, and v(t) = $\frac{5}{2}t^2 + 100$.

The position function is x(t) = $\int v(t)\,dt = \int (\frac{5}{2}t^2 + 100)\,dt = \frac{5}{6}t^3 + 100t + C$. From the initial condition x(0) = 50, x(0) = $\frac{5}{6}(0)^3 + 100(0) + C = 50$. Therefore, C = 50, and x(t) = $\frac{5}{6}t^3 + 100t + 50$.

From physics, we know that for a constant velocity, (distance traveled (d)) equals (the velocity (v)) multiplied by (a length of time (t)), that is d = vt. Calculus is necessary in order to calculate (d) over some interval of time if the velocity function v(t) is continuously varying.

Let's partition a time interval [a,b] into n equal sub-intervals $[t_{i-1}, t_i]$ between t = a and t = b, with the (n + 1) points $\{t_0, t_1, \ldots, t_n\}$ such that a = $t_0 < t_1 < t_2 < \cdot\cdot\cdot < t_{n-1} < t_n = b$. Let v($c_i$) be the velocity at any point in time c_i, where $t_{i-1} \leq c_i \leq t_i$. Then (v$(c_i) \cdot \Delta t_i$) is the area of a rectangle which is approximately equal to the distance traveled in the time interval [$t_{i-1}, t_i]$. Therefore, adding up the area of these n rectangles, written $\sum_{i=1}^{n} v(c_i)\Delta t_i$, we have an approximation to the distance traveled in the time interval [a,b]. This is a Riemann sum, and the approximation gets better as we let the number of rectangles (n) get larger. So then the exact distance traveled is $\lim_{n\to\infty} \sum_{i=1}^{n} v(c_i)\Delta t_i = \int_a^b v(t)\,dt$. Let's consider an example.

Example 2: If a mass is moving with velocity v(t) = $(3t^2 - 7t + 10)$ meters/sec. on the time interval [2, 12], find the distance traveled on this interval of time.

From above, we have d = $\int_2^{12} (3t^2 - 7t + 10)\, dt$, which is calculated

$[t^3 - \frac{7}{2}t^2 + 10t]_2^{12} = [(1728 - 504 + 120) - (8 - 14 + 20)] = 1330$ meters.

(**Work and Energy**)

Let's discuss a bit of physics in order to better understand the following example. We are all familiar with the formula F = ma, that is, the force impressed upon a mass equals the mass of the object multiplied by the acceleration it experiences. We will note that the units of force here are Newtons and the unit of distance is the meter.

Work and energy are considered to be two sides of the same coin in physics, so they are measured in the same units. Work is the expenditure of energy, and energy is the capacity to do work.
Work = (F) $\cdot$ (x) = Force multiplied by a distance, and is measured in the units of Newton-meters or Joules.
Kinetic Energy (KE) = $\frac{1}{2}mv^2$, and is also measured in Joules.

In physics, if we have a constant force multiplied through a distance, then it is easy to calculate the amount of work done, without calculus. However if we have a continuously varying force (if it varies with position) then calculus is necessary to calculate the amount of work done. As in the previous example with velocity and time, in calculus we consider very small changes and add up all of these very small changes. So we will consider a variable force F(x), dependent upon position x, over an interval of distance [a,b]. We use a set of (n + 1) points $\{x_0, x_1, \ldots, x_n\}$ to define a partition of the interval [a,b], where a = $x_0 < x_1 < x_2 < \cdot \cdot \cdot < x_{n-1} < x_n = b$. We let c_i be any point such that $x_{i-1} \le c_i \le x_i$, for each i. Then $F(c_i) \cdot (\Delta x_i)$ is the area of a rectangle where the area approximates the work done in the interval of distance $[x_{i-1}, x_i]$. Then adding up the area of all of these n rectangles gives us an approximation to the total work done on the interval of distance [a,b]. So we have a Riemann sum $\sum_{i=1}^{n} F(c_i)\Delta x_i$, and it follows that the exact amount of work done = W = $\lim_{n \to \infty} \sum_{i=1}^{n} F(c_i)\Delta x_i = \int_a^b F(x)\, dx$. This of course assumes that the work is done along a straight line. Let's consider an example.

Example 3: If a variable force acting on a mass is given by F(x) = $13x^2 + 2x$ on the interval of distance [0, 5], then the total work done on the mass is $\int_0^5 F(x)\, dx = \int_0^5 (13x^2 + 2x)\, dx$ = $[\frac{13}{3}x^3 + x^2]_0^5 \approx [(541.67 + 25) - (0)]$ = 566.67 Joules.

There is another important result in physics called the Work-Energy theorem. Using the work integral from above,

$$\int_a^b F(x)\, dx = \int_a^b ma\, dx \quad \text{(since F = ma)}$$

$$= \int_a^b m\frac{dv}{dt}\, dx \quad \text{(since } \frac{dv}{dt} = a\text{)}$$

$$= \int_a^b m\frac{dv}{dx}\frac{dx}{dt}\, dx$$

$$= \int_{v(a)}^{v(b)} mv\, dv \quad \text{(since } \frac{dx}{dt} = v, \text{ and the variable changes to v)}$$

$$= [\tfrac{1}{2}m(v(b))^2 - \tfrac{1}{2}m(v(a))^2] = \Delta KE.$$

What this says is that the work done over the distance interval [a,b] equals the change in kinetic energy over [a,b]. If we know the mass m and assume that the velocity started out at 0 at x = a, then we can determine the velocity at x = b.

(**Flux through a Surface**)

We have a pipe of cross-sectional area A. Water flows through it with a velocity v(t) which depends on time. Since the velocity is variable, we need calculus to determine the volume of water that flows through a certain cross section from time t = a to t = b. If we partition the time interval [a,b] as we have done before and if c_i is a point in the sub-interval $[t_{i-1},\ t_i]$, the product $(A \cdot v(c_i) \cdot \Delta t_i)$ is the area of a rectangle, the value of which approximates the volume of water that flows through the cross section of area A during the time interval $[t_{i-1},\ t_i]$. The sum $\sum_{i=1}^{n}(A)v(c_i)\Delta t_i$ approximates the total volume of water that flows through the cross section from time t = a to time t = b. The approximation gets better as the partition is refined and the number of rectangles increases. The exact amount of water through the cross section during the time interval [a,b] is given by the $\lim_{n \to \infty} \sum_{i=1}^{n}(A)v(c_i)\Delta t_i = \int_a^b (A)v(t)\,dt$.

Example 4: If the cross section of a water pipe is A = 4.0 cm^2, and from time t = 0 to t = 3 seconds the velocity of the water in the pipe is given by $v(t) = 12 + \frac{5}{2}t + 6t^2$ cm/sec. What volume of water passes through a cross section of the pipe in the time interval [0,3]?

The volume of water V = $\int_0^3 (A)v(t)\,dt = \int_0^3 (4)(12 + \frac{5}{2}t + 6t^2)\,dt$

$= 4[12t + \frac{5}{4}t^2 + 2t^3]_0^3 = 4(36 + \frac{45}{4} + 54) = 405\ cm^3$.

Example 5: Blood is flowing through an artery at a point of cross section

A(t) = $(2 + \frac{(0.008)}{t^2})\ cm^2$, with velocity v(t) = $4\,t^2 - \frac{1}{8}t^3$ cm/sec from time t = 1 to time t = 5 seconds. Here the rate of flow and the cross sectional area depend on time. What is the total volume of blood that flows through the particular cross section in the time interval [1,5]?

The volume is V = $\int_1^5 A(t) \cdot v(t)\ dt = \int_1^5 (2 + \frac{(0.008)}{t^2})(4t^2 - \frac{1}{8}t^3)\ dt$

$$= \int_1^5 (8t^2 - \frac{1}{4}t^3 + (0.032) - (0.001)t)\ dt$$

$$= [\frac{8}{3}t^3 - \frac{1}{16}t^4 + (0.032)t - (0.0005)t^2]_1^5$$

$$\approx (294.418) - (2.635) = (291.783)\ cm^3.$$

Example 6: Physicists measure the amount of magnetic flux Φ perpendicularly through a surface of area A in units of Webers, and the magnetic field intensity B, at a point on that surface, in units of Webers/meter^2 or Teslas. In other words, $B = \frac{\Phi}{A}$ or $\Phi = BA$. If we have a flat surface of area 2.0 meter^2 and a perpendicular magnetic field through the surface of varying intensity B(t) = (100)sin(4t) Teslas on the time interval [0,10] (in seconds), find the total amount of magnetic flux through the surface on the time interval [0,10].

The total flux (or net flux) through the surface is,

$$\Phi = \int_0^{10} B(t) \cdot A\ dt = \int_0^{10} (100) sin(4t) \cdot (2)\ dt$$

$$= (200) \int_0^{10} sin(4t)\ dt = (\frac{200}{4}) \int_0^{10} (4)\ sin(4t)\ dt$$

$$= (50) \cdot (-1) \cdot [cos(4t)]_0^{10}$$

$$= (-50)\,(cos(40) - cos(0))$$

$$\approx 11.698 \text{ Webers.}$$

EXERCISES:

Solve the following physics problems:

(1.) A mass is moving directly upward, where the only force acting on the mass is the force of gravity which provides a constant acceleration g = $-9.8 \frac{meters}{second^2}$ (ignoring air resistance). In addition, x(0) = 200 meters and v(0) = $300 \frac{meters}{second}$. The constant acceleration is negative because it acts in the downward direction. An initial position of x(0) = 200 means that the mass is being launched vertically at time t = 0, from a height of 200 meters above ground level (ground level is x = 0). The initial velocity of v(0) = 300 $\frac{meters}{second}$ is positive because it is launched vertically upward with this velocity at time t = 0.

(a) Find the acceleration, velocity, and position functions a(t), v(t), and x(t) respectively.
(b) What is the maximum height reached by the mass? (hint: the maximum height is reached when the velocity is 0)
(c) At what time will the mass hit the ground? (hint: x(t) = 0 when the mass hits the ground)
(d) With what velocity does the mass hit the ground?

(2.) An object is moving with acceleration a(t) = 3t + 6.5 $\frac{meters}{second^2}$ along the x-axis, starting at rest (v(0) = 0) and at position x(0) = 0 meters. Find the velocity and position functions v(t) and x(t) respectively, where t $\geq$ 0.

(3.) (a) How much work is done on a mass of 10 kilograms starting at rest from the origin and moving along the positive x-axis to the point x = 9 meters, where we have a variable force acting on it which is given by F(x) = $20x^2 + 3.5x$ Newtons ?

(b) Use the work energy theorem to figure out how fast the mass is moving at x = 9 meters?

(4.) Water is flowing through a pipe at a variable velocity given by v(t) = $(10 + (0.05)t^2)$ cm/sec. , and the cross section of the pipe at a point of interest is A = (3π) cm^2 . What volume of water flows through this cross section from time t = 0 to t = 2 seconds?

(#) **Arc Length**: If we have a curve of finite length as in the figure below, which is the locus of some continuous and differentiable function y(x) on an interval [a,b], we can specify a set of (n + 1) points $\{P_0, P_1, \ldots, P_n\}$ as shown in the figure.

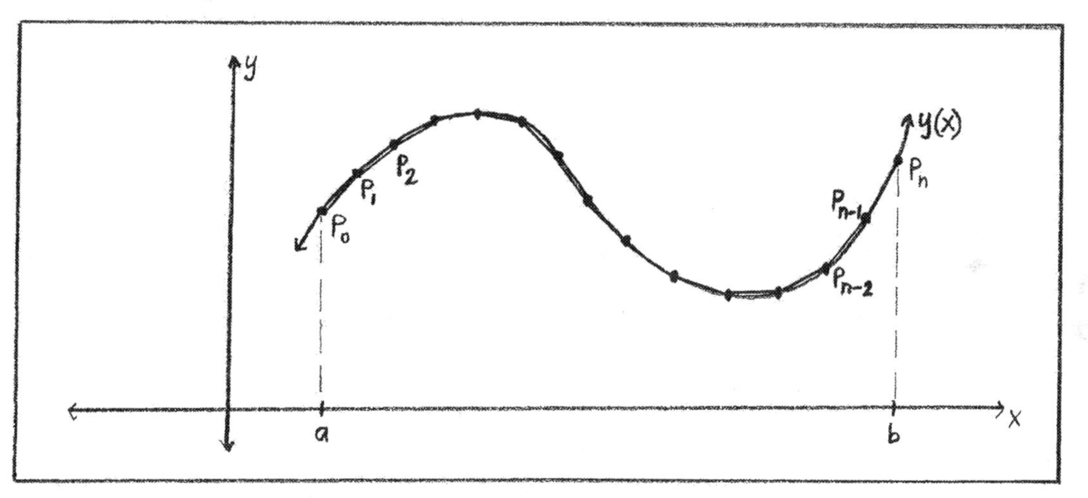

Consider the ith straight line segment from point P_{i-1} *to* P_i. The length of this segment is $\sqrt{\Delta x_i^2 + \Delta y_i^2}$ (from the pythagorean theorem) where Δx_i is the change in x and Δy_i is the change in y, from point P_{i-1} to point P_i.

Then $\sqrt{\Delta x_i^2 + \Delta y_i^2} \cdot \frac{\sqrt{\Delta x_i^2}}{\sqrt{\Delta x_i^2}} = \sqrt{1 + \left(\frac{\Delta y_i}{\Delta x_i}\right)^2} \cdot \Delta x_i$ is the length of the ith straight line segment.

From the Mean Value Theorem, we know that there is a number c_i in the ith x-subinterval corresponding to the ith arc segment such that $\left(\frac{\Delta y_i}{\Delta x_i}\right) = \frac{dy}{dx}(c_i)$, ($\frac{dy}{dx}$ evaluated at c_i). Then ($\sqrt{1+\left(\frac{dy}{dx}(c_i)\right)^2} \cdot \Delta x_i$) approximates the length of the arc from P_{i-1} to P_i.

So we have a Riemann Sum $\sum_{i=1}^{n} \sqrt{1+\left(\frac{dy}{dx}(c_i)\right)^2} \cdot \Delta x_i$ which approximates the length of the arc from x = a to x = b. The approximation gets better as $n \to \infty$, so that the exact value for the arc length L is,

$$L = \lim_{n\to\infty} \sum_{i=1}^{n} \sqrt{1+\left(\frac{dy}{dx}(c_i)\right)^2} \cdot \Delta x_i = \int_a^b \sqrt{1+\left(\frac{dy}{dx}\right)^2}\, dx .$$

Example 1: Find the length of the arc of y(x) = x^2+2, on the interval [-2,2].

$\frac{dy}{dx}$ = 2x, so the length of the arc is $L = \int_{-2}^{2} \sqrt{1+4x^2}\, dx$.

To evaluate this integral, let 2x = tan θ, x = $\frac{1}{2}tan\theta$, and dx = $\frac{1}{2}sec^2\theta\, d\theta$.

So, $\sqrt{1+4x^2} = \sqrt{1+(2x)^2} = \sqrt{1+tan^2\theta}$ = sec θ, so that $L = \frac{1}{2}\int_{x=-2}^{x=2} sec^3\theta\, d\theta$.

So we need to find an antiderivative for $sec^3\theta$. Let's use integration by parts. Let u = sec θ, du = $sec\theta tan\theta d\theta$, dv = $sec^2\theta\, d\theta$, v = $tan\theta$.

Then $\int sec^3\theta\, d\theta = sec\theta tan\theta - \int sec\theta tan^2\theta\, d\theta$

$$= sec\theta tan\theta - \int sec\theta(sec^2\theta - 1)\, d\theta$$

$$= sec\theta tan\theta - \int sec^3\theta\, d\theta + \int sec\theta\, d\theta$$

$$= sec\theta tan\theta - \int sec^3\theta\, d\theta + ln|sec\theta + tan\theta| + C$$

Therefore, $2\int sec^3\theta \, d\theta = sec\theta tan\theta + ln|sec\theta + tan\theta| + C$

So, $\int sec^3\theta \, d\theta = \frac{1}{2}sec\theta tan\theta + \frac{1}{2}ln|sec\theta + tan\theta| + C$.

Since $2x = tan\theta$, the following diagram shows that $sec\theta = \sqrt{1+4x^2}$,

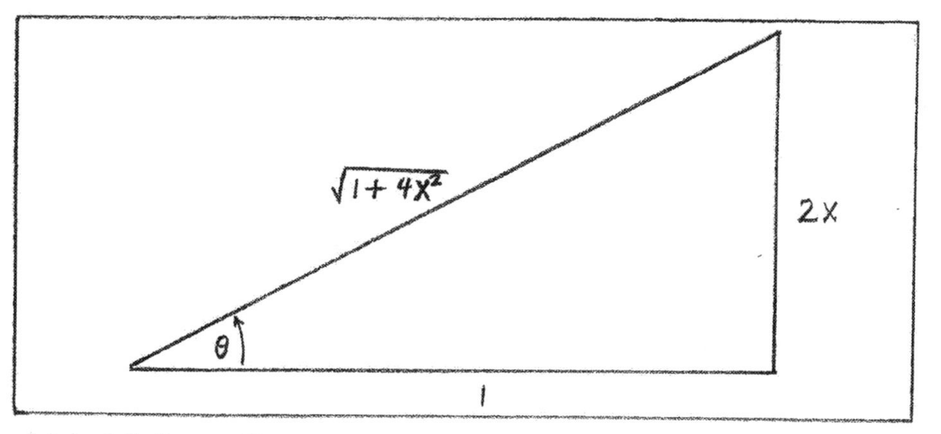

from which it follows that,

$\int sec^3\theta \, d\theta = (x)\sqrt{1+4x^2} + \frac{1}{2}ln|\sqrt{1+4x^2} + 2x| + C$.

So the arc length L = $(\frac{1}{2})\ [(x)\sqrt{1+4x^2} + \frac{1}{2}ln|\sqrt{1+4x^2} + 2x|]_{-2}^{2}$

= $(\frac{1}{2})\ [(2\sqrt{17} + \frac{1}{2}ln|\sqrt{17}+4|) \ - \ ((-2\sqrt{17}) + \frac{1}{2}ln|\sqrt{17}-4|)]$

$\approx (9.2936)$.

Example 2: What is the length of the curve defined by y = mx + b from x = 0 to x = N?

$\frac{dy}{dx} = m$, so $(\frac{dy}{dx})^2 = m^2$.

Then the arc length L = $\int_0^N \sqrt{1 + (\frac{dy}{dx})^2}\, dx \ = \ \int_0^N \sqrt{1+m^2}\, dx$

= $(\sqrt{1+m^2})\ [x]_0^N$ = (N)($\sqrt{1+m^2}$).

This arc length could have been determined by elementary methods from high school geometry. The reader should verify this for themselves as a fun exercise.

EXERCISES:

(1.) Find the length of the arc of $y = ln(sec\theta)$ from $\theta = 0$ to $\theta = \frac{\pi}{3}$.

(2.) Find the length of the arc of $y = \sqrt{1+x}$ from x = -1 to x = 8.

(3.) Find the length of the arc of $y = 4x^2 + 3$ from x = 0 to x = $\frac{1}{8}$.

(#) Surfaces of Revolution: If we have a curve in the x-y plane as in the previous section, and we were to revolve it around the x-axis, then we would have what we call a surface of revolution. We are interested in calculating its area. See the figure below. Consider one small straight line segment between points P_{i-1} and P_i .

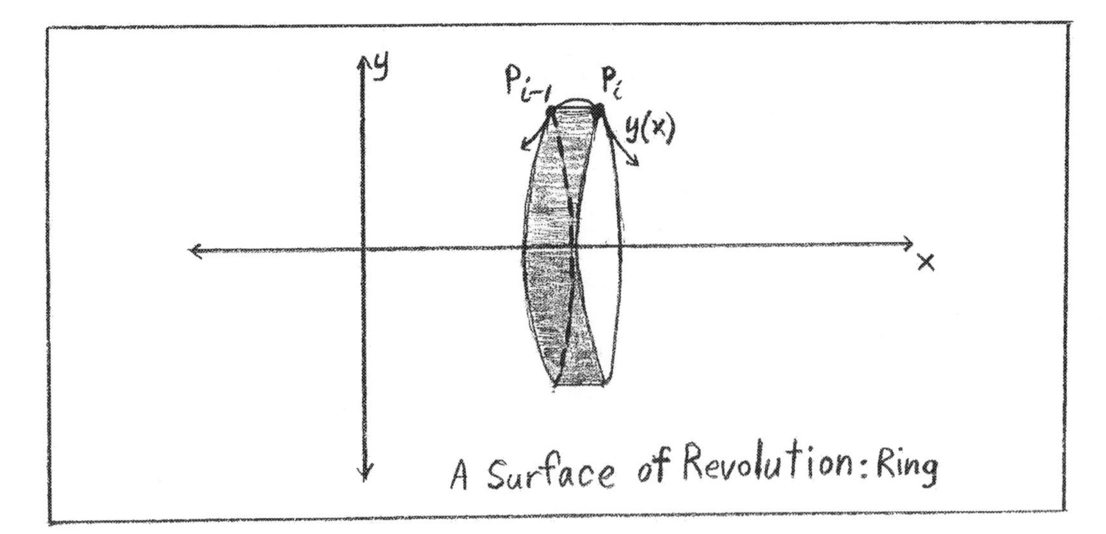

A Surface of Revolution: Ring

If we revolve just that one small straight line segment around the x-axis, then we would have a ring-like surface around the x-axis. It would have surface area approximately equal to

$$[2\pi y(c_i) \cdot \sqrt{1 + (\tfrac{dy}{dx}(c_i))}^2 \cdot \Delta x_i] .$$

The new part "$2\pi y(c_i)$" is the circumference of the revolution and we multiply it by the length of the segment to get a surface area.

If we revolved all n of the segments (on the curve over interval [a,b]), in the same way, then we would have a segmented surface with total area approximating the exact area that would be obtained by revolving the actual curve over [a,b] around the x-axis. So we would calculate the area of the surface of revolution as the

$$\lim_{n\to\infty} \sum_{i=1}^{n} [(2\pi y(c_i) \cdot \sqrt{1 + (\tfrac{dy}{dx}(c_i))^2} \cdot \Delta x_i] = \int_a^b 2\pi y(x)\sqrt{1 + (\tfrac{dy}{dx})^2}\, dx .$$

Example 1: Let y(x) = r (a constant) over the interval [a,b]. If we revolved that horizontal line segment around the x-axis we would have a pipe-like surface of radius r. This is shown in the figure below.

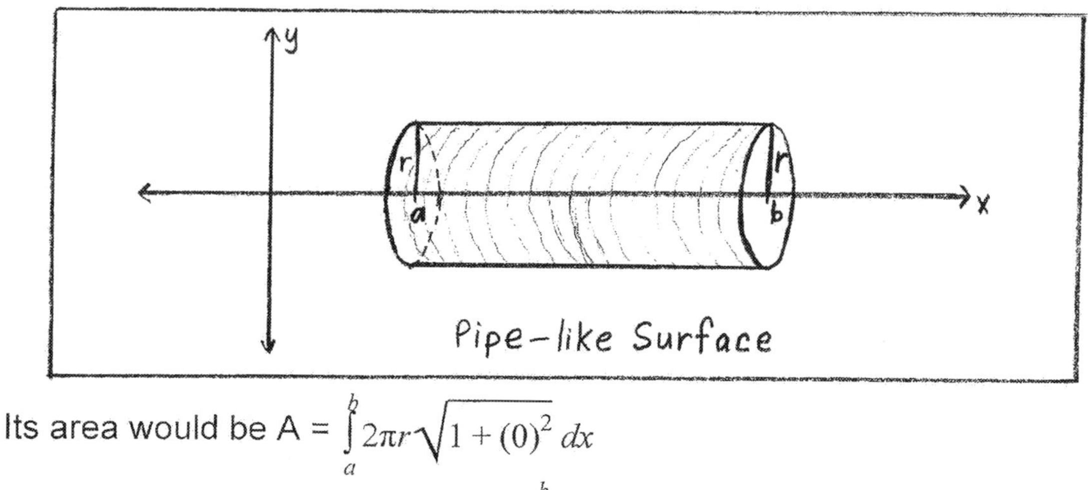

Its area would be A = $\int_a^b 2\pi r\sqrt{1 + (0)^2}\, dx$

(since $\frac{dy}{dx}$ would be 0 on [a,b]) = $\int_a^b 2\pi r\, dx = (2\pi r) \cdot [x]_a^b = 2\pi r(b - a)$.

Example 2: Let y(x) = $\sqrt{x}$ on the interval [0,9]. Revolve this curve about the x-axis. This situation is drawn in the figure below.

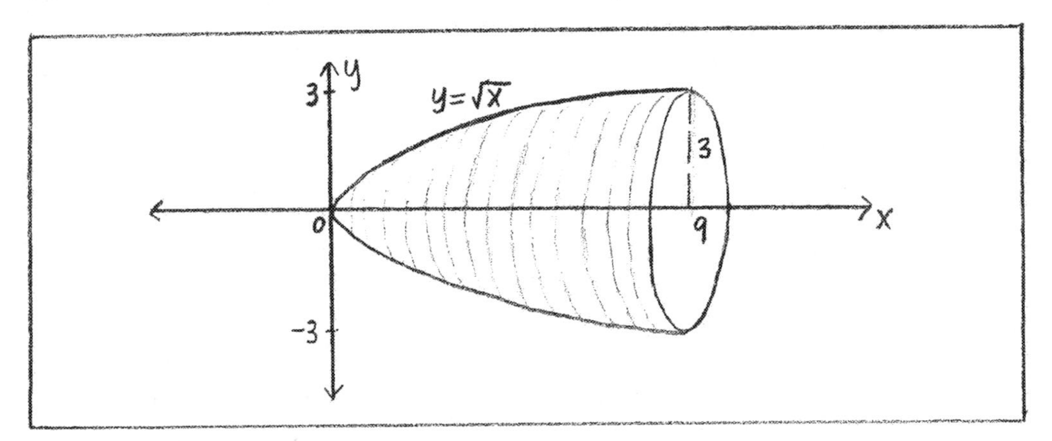

The area of the resulting surface of revolution would be

$$A = \int_0^9 2\pi\sqrt{x} \cdot \sqrt{1 + (\tfrac{1}{2\sqrt{x}})^2}\, dx = \int_0^9 2\pi\sqrt{x} \cdot \sqrt{1 + \tfrac{1}{4x}}\, dx$$

$$= \int_0^9 2\pi\sqrt{x + \tfrac{1}{4}}\, dx$$

$$= (2\pi)(\tfrac{2}{3}) \cdot [(x + \tfrac{1}{4})^{\frac{3}{2}}]_0^9$$

$$= \tfrac{4\pi}{3}((9.25)^{\frac{3}{2}} - (0.25)^{\frac{3}{2}}) \approx (117.3187).$$

Example 3: Let y(x) = $\frac{1}{x}$, on the interval [1, ∞). If we revolve this curve around the x-axis, then we get an infinitely long (from 1 to ∞) funnel-like surface of revolution, shown in the figure below. Let's calculate its surface area. It would be,

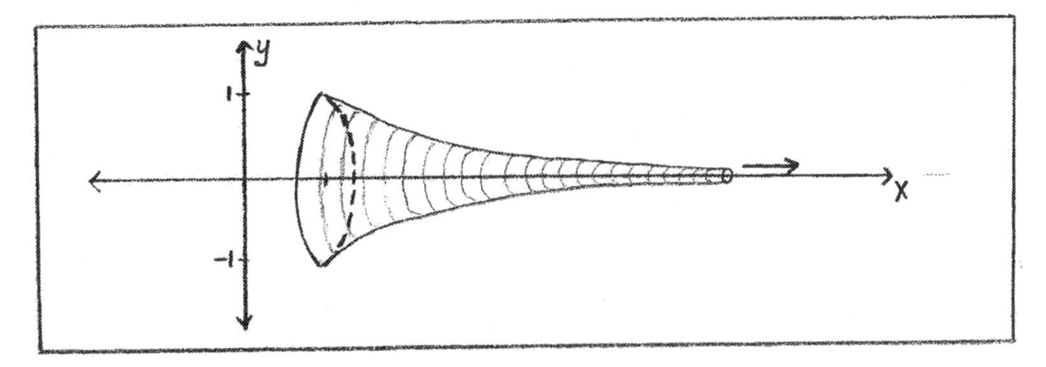

$A = \int_1^{\infty} 2\pi(\frac{1}{x}) \cdot \sqrt{1 + (\frac{1}{x^4})}\, dx$, because the $\frac{d}{dx}(\frac{1}{x}) = (\frac{-1}{x^2})$, so $\left(\frac{dy}{dx}\right)^2 = \frac{1}{x^4}$.

We are using an integral with ∞ as the upper limit, which is a type of integral which we call an improper integral. Our use of them in this example and in Example 1 in the next section on Solids of Revolution should cause no difficulties in understanding the results. All types of Improper integrals will be fully and correctly treated in chapter (11).

Now since $(\frac{1}{x})\sqrt{1 + (\frac{1}{x^4})} > (\frac{1}{x})$ for all x in the interval $[1, \infty)$, we can see that

$$A = \int_1^{\infty} 2\pi(\frac{1}{x}) \cdot \sqrt{1 + (\frac{1}{x^4})}\, dx > \int_1^{\infty} 2\pi(\frac{1}{x})\, dx$$

$$= (2\pi) \cdot [ln|x|]_1^{\infty}$$

$$= (2\pi)(ln(\infty) - 0) = \infty$$

Therefore, the surface area of this infinitely long funnel is ∞. This is perhaps not a very surprising result if it weren't for the fact that the volume enclosed by this funnel from 1 to ∞ is finite! We'll show this in the next section on solids of revolution.

EXERCISES:

(1.) Find the area of the surface generated by revolving the line y = 3x, from x = 0 to x = 10, about the x-axis.

(2.) Find the area of the surface generated by revolving the curve y = e^{-x}, from x = 0 to x = 10, about the x-axis.

(3.) Find the surface area generated by revolving $y = \sqrt{r^2 - x^2}$, from x = -r to x = r, about the x-axis.

(#) **Solids of Revolution**: Now we will consider the volume enclosed by a surface of revolution. The more usual way of thinking about this problem is to imagine the area under a curve on some interval of the x-axis being revolved about the x-axis, and the volume that would result.

(**Volumes using Discs**)

For the non-negative function y(x), continuous on the interval [a,b], we consider the graph of this in the x-y plane and a typical rectangle extending from the x-axis (its base is on the x-axis) upward to the curve y(x). With the partition that we make on the interval of integration [a,b], the base of the rectangle is one of the partition intervals $[x_{i-1}, x_i]$, and the height of the rectangle is y(c_i) where c_i is any number in the sub-interval $[x_{i-1}, x_i]$. This situation is shown in the figure below.

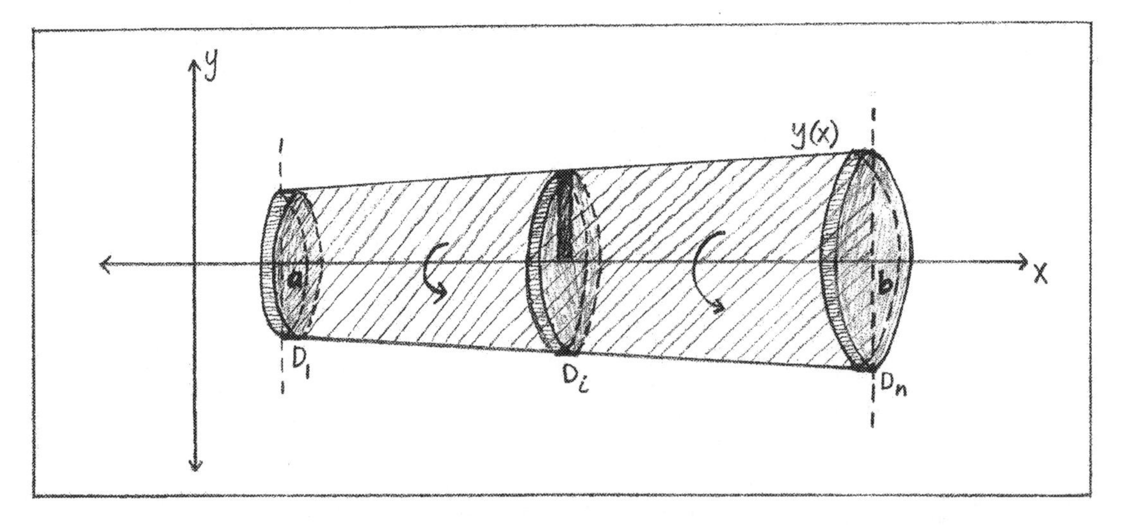

Now, imagine the rectangle being revolved around the x-axis with its base fixed to the x-axis. The resulting solid is a disc with width $\Delta x_i = (x_i - x_{i-1})$ and radius y(c_i). The volume of the disc is $D_i = \pi(y(c_i))^2 \cdot (\Delta x_i)$. The volume of the resulting solid of revolution on the entire interval [a,b] is approximated by adding up the volumes of the n discs $\{D_i\}_{i=1}^{n}$. The exact volume of the solid of revolution is given by the following,

$$V = \lim_{n\to\infty} \sum_{i=1}^{n} D_i = \lim_{n\to\infty} \sum_{i=1}^{n} \pi(y(c_i))^2 \cdot (\Delta x_i) = \int_a^b \pi(y(x))^2 \, dx .$$

We will consider a case as our first example, where the upper limit will be ∞. As we discussed in Example 3 in the Surfaces of Revolution section, this should present no difficulties in understanding before we correctly treat integrals like these (Improper Integrals) in chapter (11). The analysis is exactly the same, but we have a partition P extending over an interval of infinite length. The volume of the solid of revolution will be made up from a sum of an infinite number of discs, as usual, but we state things differently in this case. Let the widest of the $\{\Delta x_i\}_{i=1}^{\infty}$ be called the norm of the partition, denoted ||P|| . Then the exact volume is

$$V = \lim_{\|P\|\to 0} \sum_i \pi(y(c_i))^2 \Delta x_i = \int_a^\infty \pi(y(x))^2 \, dx \text{, provided the limit exists.}$$

(**<u>Volumes using Washers</u>**)

This is a similar situation, but we have two curves $y_1(x)$ and $y_2(x)$ continuous on an interval [a,b], such that $0 \le y_1(x) \le y_2(x)$ on [a,b]. This situation is shown below,

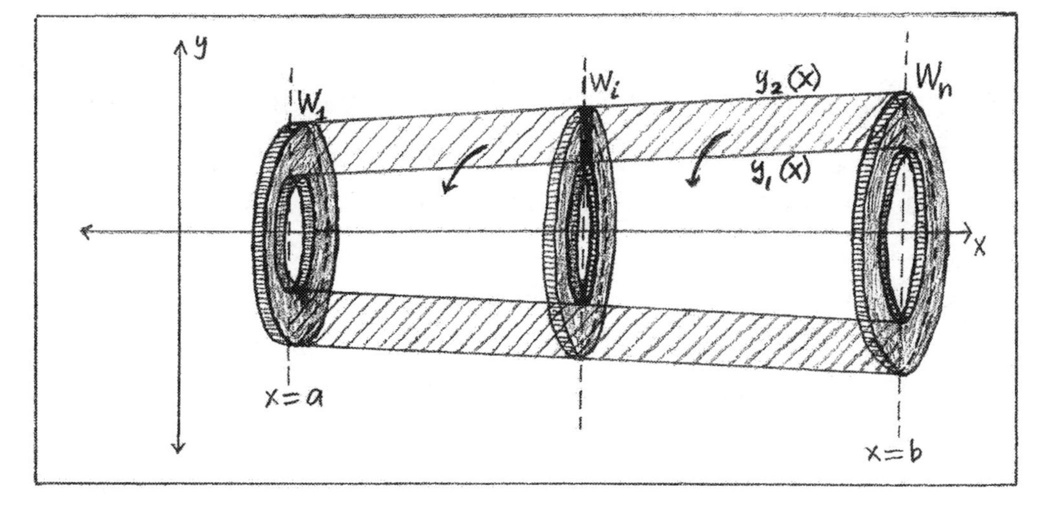

The partition of the interval [a,b] is as before, but the rectangle extends from the lower function to the upper function. The lower function plays the role of the x-axis. So the height of the rectangle in the ith part of the partition is $(y_2(c_i) - y_1(c_i))$, where $x_{i-1} \le c_i \le x_i$. When this type of rectangle is revolved about the x-axis, the result is a solid like a washer (it has a hole in the middle). So the volume for the solid of revolution in this case is approximated by adding up the volumes of all the washers on [a,b].

Then the exact volume is the $\lim_{n\to\infty} \sum_{i=1}^{n} (\pi(y_2(c_i))^2 - \pi(y_1(c_i))^2) \cdot \Delta x_i$

which will converge to $\int_a^b (\pi(y_2(x))^2 - \pi(y_1(x))^2)\, dx$.

Example 1: Revolve the area under the function y(x) = $\frac{1}{x}$ on [1, ∞) about the x-axis. This will be an infinitely long solid of revolution. We showed previously that the surface of this solid is a funnel-like surface with infinite surface area. We would like to show that the volume of the corresponding solid of revolution is finite. Using the analysis above for a volume derived from discs, we can see that the volume is,

V = $\int_1^{\infty} \pi(\frac{1}{x})^2\, dx = \pi \int_1^{\infty} x^{-2}\, dx$

= $(\pi) \cdot [\frac{-1}{x}]_1^{\infty}$

= $(\pi) \cdot (0 - (-1))$

= (π).

Example 2: If we take the area between the curves $y_1(x) = \sqrt{2x}$, and $y_2 = \sqrt{3x}$, over the interval [1,10], and revolve it about the x-axis, what would be the volume of the resulting solid of revolution?

This is a simple problem where we will be adding up the volumes of an infinite number of washers, which is sort of what we are actually doing in an integration problem. The integral for computing the exact volume is V =

$(\pi) \int_1^{10} [(y_2(x))^2 - (y_1(x)^2]\, dx$

$$= (\pi)\int_1^{10} [(\sqrt{3x})^2 - (\sqrt{2x})^2]\, dx$$

$$= (\pi)\int_1^{10} (x)\, dx = (\pi)\cdot [\tfrac{x^2}{2}]_1^{10}$$

$$= (49.5)(\pi)\,.$$

Example 3: Prove that the volume of a right circular cone, with base radius r and height h, is $\frac{1}{3}\pi r^2 h$. If we look at the cone from a side view as in the figure below,

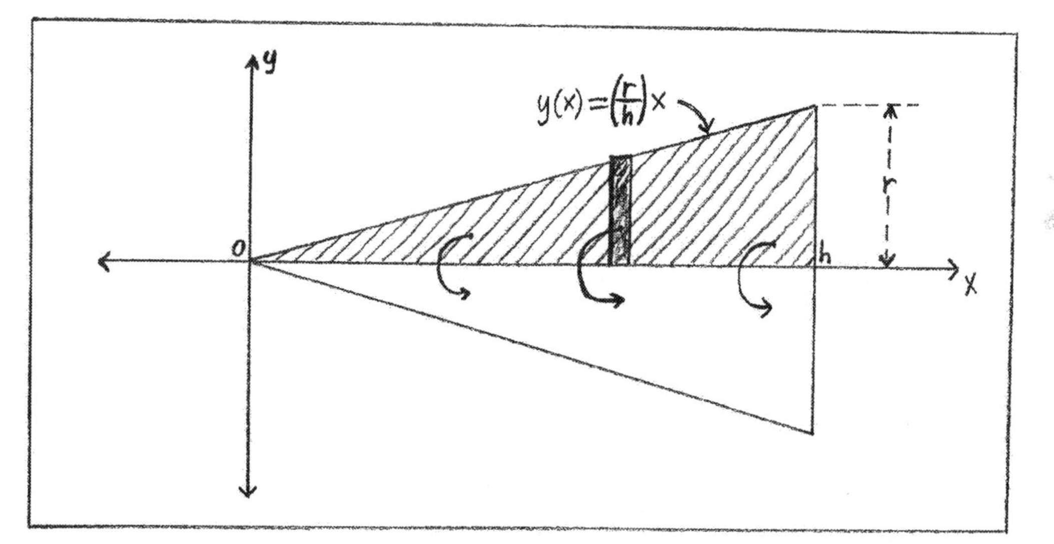

where we put the top most point at the origin and the axis of the cone is along the x-axis from the origin to the point h, then it is easy to find the volume. We just revolve about the x-axis the region that is within the lines $y(x) = (\frac{r}{h})x$, x = h, and the x-axis. A typical rectangle is shown in the figure.

$$V = (\pi)\int_0^h (y(x))^2\, dx = (\pi)\int_0^h \tfrac{r^2}{h^2}x^2\, dx$$

$$= (\pi)\tfrac{r^2}{h^2}[\tfrac{x^3}{3}]_0^h = (\pi)\tfrac{r^2}{h^2}(\tfrac{h^3}{3} - 0)$$

$$= \tfrac{1}{3}\pi r^2 h$$

EXERCISES:

(1.) Find the volume generated by revolving the area under $y = \sqrt{r^2 - x^2}$ and above the x-axis, from x = -r to x = r , about the x-axis.

(2.) Find the volume generated by revolving the area under y = x^2 and above the x-axis, from x = 0 to x = 1, about the x-axis.

(3.) Find the volume generated by revolving the area between $y = \sqrt[3]{x}$ and the x-axis, from x = -1 to x = 1, about the x-axis.

(11) Improper Integrals

(#) Improper Integrals of Two Types: So far we have been dealing with (except for a couple of integrals in the examples of the last chapter) definite integrals where the limits of integration have been finite numbers and where the integrand has been continuous and bounded on the interval of integration. We now want to consider cases where these conditions are not satisfied. These types of integrals we call improper integrals. There are two main cases to consider:

(A) An infinite limit, such as in $\int_2^{\infty} y(x)\,dx$ or $\int_{-\infty}^{-1} y(x)\,dx$.

(B) An integrand that becomes unbounded at one of the limits, such as in $\int_0^1 \frac{1}{\sqrt{x}}\,dx$ or $\int_1^6 \frac{7}{6-x}\,dx$. The function $\frac{1}{\sqrt{x}}$ approaches ∞ as we approach 0 from the right. The function $\frac{7}{6-x}$ approaches ∞ as we approach 6 from the left. Many times integrals like these may correspond to areas which are actually finite, but at other times the corresponding area is infinite.

Let's consider integrals of the first type, which have an infinite limit.

Example 1: Evaluate the integral $\int_1^{\infty} \frac{1}{x^2}\,dx$. Check the figure below.

Note that this can be written $\int_1^{\infty} x^{-2}\,dx$. The correct way to evaluate an integral like this is to let the upper limit be a variable like s, so that we have $\int_1^{\infty} x^{-2}\,\mathrm{dx} = \lim_{s\to\infty} \int_1^{s} x^{-2}\,dx = \lim_{s\to\infty} [\frac{-1}{x}]_1^s = \lim_{s\to\infty}(\frac{-1}{s} - (-1)) = \lim_{s\to\infty}(1 - \frac{1}{s}) = 1$.

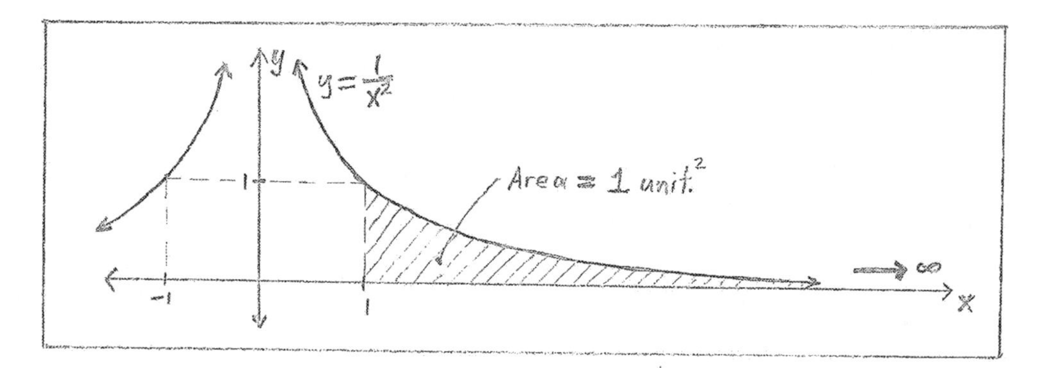

So this stretch of area underneath the curve y = $\frac{1}{x^2}$ from 1 to ∞ is finite in extent. The reason that it is finite is because the curve $\frac{1}{x^2}$ approaches the x-axis very fast for x > 1. For many functions that approach the x-axis at a slower rate, the integral would be infinite.

Usually, we do a shorthand kind of thing by writing:
$\int_1^{\infty} x^{-2}\,dx = [\frac{-1}{x}]_1^{\infty} = ((\frac{-1}{\infty}) - (-1)) = (1 - \frac{1}{\infty}) = 1$. When we do this we are treating ∞ as if it were a number, which it is not. However, we understand that $(\frac{1}{\infty})$ actually means (1 divided by a quantity which is approaching $+\infty$). Such a ratio is approaching zero.

<u>Example 2</u>: Evaluate $\int_{-\infty}^{-2} e^x\,dx$. The reader can refer to the graphs of common functions in Chapter (3) of this book for the graph of $y = e^x$.

So, $\int_{-\infty}^{-2} e^x\,dx = \lim_{s \to -\infty} \int_s^{-2} e^x\,dx = \lim_{s \to -\infty} [e^x]_s^{-2} = \lim_{s \to -\infty} (e^{-2} - e^s)$

$= (\frac{1}{e^2}) - (0) = \frac{1}{e^2}$.

Once again an infinite stretch of area turns out to be finite in extent.

<u>Example 3</u>: Evaluate $\int_1^{\infty} \frac{1}{1+x^2}\,dx$.

$\int_1^{\infty} \frac{1}{1+x^2}\,dx = \lim_{s \to \infty} \int_1^{s} \frac{1}{1+x^2}\,dx = \lim_{s \to \infty} [tan^{-1}(x)]_1^s$

$$= \lim_{s \to \infty} (tan^{-1}(s) - tan^{-1}(1)) = (\tfrac{\pi}{2} - \tfrac{\pi}{4}) = \tfrac{\pi}{4}.$$

Note that since $\frac{1}{1+x^2} < \frac{1}{x^2}$ is true for all x > 1, it is not surprising that $\int_1^{\infty} \frac{1}{1+x^2}\,dx < \int_1^{\infty} \frac{1}{x^2}\,dx$. We are referring to the integral of Example 1.

Example 4: We will now revisit the integral that we calculated in example 3 of the section on Surfaces of Revolution. When we calculated this integral before, we used the shorthand method discussed above because we did not yet know the correct way. It was the integral $\int_1^{\infty} \frac{1}{x}\,dx$.

$$\int_1^{\infty} \frac{1}{x}\,dx = \lim_{s \to \infty} \int_1^{s} \frac{1}{x}\,dx = \lim_{s \to \infty} [ln|x|]_1^s = \lim_{s \to \infty} (ln|s| - ln|1|)$$
$$= (\infty - 0) = \infty.$$

This agrees with our previous result. This function does not approach the x-axis, for x > 1, fast enough to prevent the area under the curve from accumulating beyond all bounds.

Now, for improper integrals of the second type, where the integrand becomes unbounded in the interval of integration.

Example 5: Consider the function y(x) = $(\frac{1}{x-6})$ on [0,6]. This curve is shown in the figure below. y(x) becomes unbounded as we approach 6 from the left. It has a vertical asymptote there. If we want to find $\int_0^6 (\frac{1}{x-6})\,dx$, the correct way of doing this is:

$$\int_0^6 (\tfrac{1}{x-6})\,dx = \lim_{s \to 6^-} \int_0^s (\tfrac{1}{x-6})\,dx = \lim_{s \to 6^-} [ln|x-6|]_0^s$$
$$= \lim_{s \to 6^-} (ln|s-6| - ln|-6|)$$
$$= ((-\infty) - ln(6)) = -\infty.$$

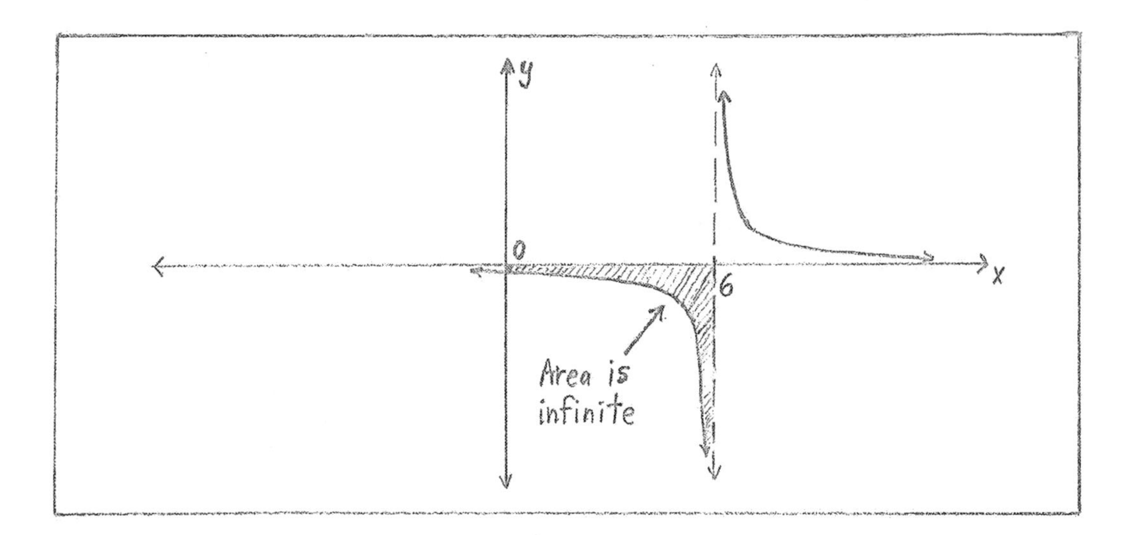

The area corresponding to this integral is actually the absolute value of this result, which would be $(+\infty)$. It is infinite in extent.

Example 6: Now consider the function y(x) = $(\frac{1}{\sqrt{x-6}})$ on the interval [6,10]. The graph of this function is shown in the figure below. Note that its domain is (6, ∞).

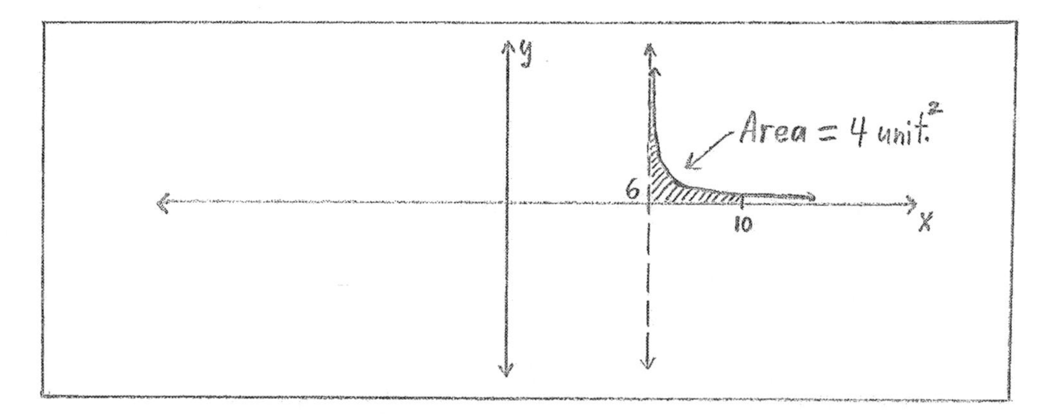

Let's find the integral $\int_6^{10} \frac{1}{\sqrt{x-6}}\, dx$.

$$\int_6^{10} \frac{1}{\sqrt{x-6}}\, dx = \lim_{s\to 6^+} \int_s^{10} (x-6)^{-\frac{1}{2}}\, dx = \lim_{s\to 6^+} [2\sqrt{x-6}\,]_s^{10}$$

$$= \lim_{s\to 6^+} (2\sqrt{4} - 2\sqrt{s-6}\,)$$

$$= (4-0) = 4.$$

For this function on [6,10], the area corresponding to the integral is finite in extent, despite the fact that the integrand approaches $(+\infty)$ as x approaches 6 from the right. When we compare the situations in Example 5 and in this example, we recognize it must be true that the function of this example approaches its vertical asymptote much more quickly, so that the area here turns out to be finite.

EXERCISES:

Evaluate the following integrals.

(1.) $\int_0^1 \frac{1}{x-1}\,dx$

(2.) $\int_1^3 \frac{1}{\sqrt[3]{x-1}}\,dx$

(3.) $\int_0^\infty e^{-4x}\,dx$

(4.) $\int_5^\infty \frac{1}{x^3}\,dx$

(5.) $\int_1^\infty \frac{1}{3\sqrt{x}}\,dx$

Part IV

Differential Equations

(12) First Order Differential Equations

(#) Introduction: Differential Equations is one of the most important areas of mathematics, especially in the physical sciences and engineering. It is a direct outgrowth of differential and integral calculus. In this final part of the book, we will go over some of the most common and useful types of ordinary differential equations (ODE's), along with some applications. ODE's are equations involving ordinary derivatives and differentials, which is what we have been studying in this book where we have discussed functions of a single variable. There are also partial differential equations (PDE's) involving so-called partial derivatives. We have not studied partial derivatives because that is a topic in the calculus of functions of several variables. PDE's are important also, but are beyond the scope of this book. The differential equations that we will study here should give the reader a good introduction to the main concepts of this vast subject. Here are three examples of an ordinary differential equation:

(A) $2\frac{dy}{dx} + x^2y = e^x$

(B) $\frac{d^2y}{dx^2} + 3x\frac{dy}{dx} + 4xy = 2x$

(C) $(x^3y + 4y^2)\,dx + (xy^3 - 5x^2)\,dy = 0$

When we have an ODE involving only the first derivative of a function, we call the ODE a first order differential equation. Second order ODE's involve derivatives up to the second order, and so on. Up to this point in mathematics, the student has been solving equations where the solution is a set of numbers. In the subject of differential equations, the solution of the equation is some function y(x) or relation f(x,y) that satisfies some relationship involving derivatives and differentials.

(#) Variables Separable Equations: This is usually the simplest type of ODE. It is an ODE where the two variables can be separated and the solution is found by integration. Let's consider some examples.

Example 1: Solve the differential equation $\frac{dy}{dx} = \frac{xe^{ax^2}}{y^2 + y}$.
Here the variables can be separated by multiplying through by $(y^2 + y)$ and by (dx), to get,
$(y^2 + y)\, dy = xe^{ax^2}\, dx$.

So, $\int (y^2 + y)\, dy = \int xe^{ax^2}\, dx$.

So, $\frac{y^3}{3} + \frac{y^2}{2} + C_1 = (\frac{1}{2a})e^{ax^2} + C_2$.

We can write the solution after combining constants and re-arranging,
$\frac{y^3}{3} + \frac{y^2}{2} - (\frac{1}{2a})e^{ax^2} = C$

This is a relation involving x and y which satisfies the ODE. If we had an initial condition, then we could solve for C.

Example 2: Solve $\frac{1}{mx^3}\, dy = \frac{1}{y^2}\, dx$, with the initial condition y(0) = 3.
Getting rid of fractions and separating the variables, we have,
$y^2\, dy = mx^3\, dx$. Integrating both sides,

$\int y^2\, dy = \int mx^3\, dx$, leads to $\frac{y^3}{3} = \frac{mx^4}{4} + C$, or,

$y = \sqrt[3]{\frac{3mx^4}{4} + C}$. Using the initial condition, we have C = 27.
Therefore, the solution to this initial value problem is,
$y = \sqrt[3]{\frac{3mx^4}{4} + 27}$.

Example 3: Solve $\frac{dy}{dx} + cot^2(x)\frac{dy}{dx} = y^2$, with initial condition y(0) = $\frac{\pi}{4}$.

Now, we have $(1+cot^2(x))\frac{dy}{dx}=y^2$.

Then, using a trigonometric identity, we have
$(csc^2(x))\frac{dy}{dx}=y^2$.

Then, separating the variables, we have,
$\frac{1}{y^2}\,dy = sin^2(x)\,dx$. We can integrate both sides,

$\int\frac{1}{y^2}\,dy=\int sin^2(x)\,dx$. Use the trigonometric identity: $sin^2(x)=\frac{1-cos(2x)}{2}$,

So, $\int\frac{1}{y^2}\,dy=\frac{-1}{y}=\int\frac{1-cos(2x)}{2}\,dx \quad\Rightarrow\quad \frac{-1}{y}=\frac{1}{2}\int dx-\frac{1}{2}\int cos(2x)\,dx$

$\Rightarrow \quad \frac{-1}{y}=\frac{1}{2}x-\frac{1}{4}sin(2x)+C$. Then solving for y,
$y=\frac{1}{\frac{1}{4}sin(2x)-\frac{1}{2}x+C}$.

From the initial condition, $\frac{\pi}{4}=\frac{1}{C}$, or C = $\frac{4}{\pi}$.
Therefore, we have our final solution:
$y=\frac{1}{\frac{1}{4}sin(2x)-\frac{1}{2}x+\frac{4}{\pi}}$.

Example 4: There are phenomena in nature that can be modeled well with the following first order separable differential equation:

$$\frac{dP}{dt}=kP.$$

The dependent variable P is actually a function of time P(t), and k is a constant. What this says is that the rate at which P(t) is changing is proportional to the magnitude of P(t) itself. The larger P(t) is, the greater the rate of change $\frac{dP}{dt}$. The smaller P(t) is, the lesser is the rate of change $\frac{dP}{dt}$. This makes sense in many situations such as human population growth (over some periods of time in some places, as we'll see in the next example), or in the growth of a bank account that accrues interest compounded continuously . A bank account accumulating interest compounded continuously will grow faster if there is more money in the account.

If k is positive, then this equation corresponds to some kind of exponential growth. If k is negative, then this equation corresponds to some kind of exponential decay. Let's solve this differential equation, along with some initial conditions to prove these assertions.

$\frac{dP}{dt} = kP \quad \Rightarrow \quad \frac{dP}{P} = k\,dt$.

Then $\int \frac{dP}{P} = \int k\,dt \quad \Rightarrow \quad ln(P) = kt + C$

$\Rightarrow \quad e^{ln(P)} = e^{kt+C} \quad \Rightarrow \quad P(t) = e^{C} \cdot e^{kt}$

(e^{C}) is a positive constant, which we can simply call P_0, which is the value of P(t) at time t = 0.

Therefore, our solution is: $P(t) = P_0 e^{kt}$, for $t \geq 0$.

This shows that P(t) is a function involving exponential change, which depends on the starting value for P(t) at time t = 0, and the value of k. We need two conditions to determine the two constants P_0 and k.

(**Exponential Growth**)

Let's determine the full model for P(t) in a specific application involving growth, by specifying two conditions that enable us to find P_0 and k.

Suppose we have a bacterial culture with P(0) = P_0 = 3,000,000 bacteria at time t = 0 (suppose t is time in days). Further suppose that we have discovered that P(5) = 5,673,000 bacteria.

So, P(0) = $P_0 \cdot e^{k(0)}$ = $P_0(1)$ = P_0 = 3,000,000.
So we have P(t) = (3,000,000) $\cdot\, e^{kt}$. Now we just have to determine k.

Using the second condition, P(5) = 5,673,000 = (3,000,000) e^{5k}
$\Rightarrow \frac{5{,}673{,}000}{3{,}000{,}000} = 1.891 = e^{5k}$ $\Rightarrow$ ln(1.891) = ln(e^{5k}) = 5k.
$\Rightarrow$ k = $\frac{ln(1.891)}{5}$ = 0.1274 .

So our full model is P(t) = (3,000,000) $e^{(0.1274)t}$.

We may now find things like what is the projected bacterial count at t = 10 days, assuming that this is a good model of the situation?
We would find this simply by computing:
P(10) = (3,000,000) $e^{(0.1274)(10)}$ = 10,725,373 bacteria.

(**Exponential Decay**)

Suppose we have P(0) = P_0 = 500 grams of a radioactive element at time t = 0 (t represents time in years), and we determine that P(50,000) = 412.11 grams.

Using the first condition, $P(t) = 500e^{kt}$.

Using the second condition, P(50,000) = 412.11 = (500) $e^{(50{,}000)k}$,
$\Rightarrow$ $\frac{412.11}{500}$ = $e^{(50{,}000)k}$ $\Rightarrow$ (0.82422) = $e^{(50{,}000)k}$
$\Rightarrow$ ln(0.82422) = ln($e^{(50{,}000)k}$) = (50,000)k
So, k = $\frac{ln(0.82422)}{50{,}000}$ = -(0.000003866)

So we model the number of grams P(t) of the amount of the radioactive element remaining after t years as: P(t) = (500) $e^{-(0.000003866)t}$.

We can now determine how many grams of this radioactive element should be left after a million years? We can compute this to be approximately P(1,000,000) = (500) $e^{-(0.000003866)(1,000,000)} = 10.47$ grams.

What is the half-life (τ) of this element? The half-life of a radioactive element is the amount of time for half of the element that currently exists to decay. That would mean P(τ) = $\frac{1}{2}P_0$.
This means that we solve the equation $e^{-(0.000003866)\tau} = \frac{1}{2}$ for τ. This leads us to the result: $\tau = \frac{ln(\frac{1}{2})}{-(0.000003866)} = 179,293$ years.

Example 5: There is a significantly different and surely more realistic way (in many situations) to model population growth than with the exponential model of the previous example. Particularly, many biologists and ecologists have developed the so-called logistic differential equation in order to more realistically and adequately model the growth of animal populations, which of course includes humans on this planet earth with its finite resources. The resources that make a difference to biological organisms are such things as the amount of space, food, water, environmental quality, and so on. In ecosystems, there is a finite carrying capacity for a population, which we can call M. Certainly, the number of humans that could inhabit the earth is not unbounded. We will derive an equation for human populations where we can assume that there is a finite carrying capacity M. We will assume that there was a time that we can designate as time t = 0 when the population P(t) was at a very low level which we'll call P_0, and the carrying capacity M was not a major factor in human population growth. (Mathematically speaking, we should include this example of a differential equation since it allows us to use the technique of partial fractions in its solution). The logistic differential equation for population growth is:

$\frac{dP}{dt} = kP(M-P)$, for t $\geq$ 0.

This equation takes into account not only the population level P(t), but also the diminishing difference between M and P(t). It is variables separable in the following way: $\frac{dP}{P(M-P)} = k\,dt$. Then integrating both sides, we have

$\int \frac{dP}{P(M-P)} = \int k\,dt$. The partial fractions decomposition for the left side is

$\frac{1}{P(M-P)} = \frac{A}{P} + \frac{B}{(M-P)}$, where the constants A and B are to be determined.

So, $1 = A(M-P) + B(P)$ after multiplying through by $P(M-P)$.
Letting P = M ⇒ 1 = BM ⇒ $B = \frac{1}{M}$.
Letting P = 0 ⇒ 1 = AM ⇒ $A = \frac{1}{M}$.

So that $\int \frac{1}{P(M-P)} dP = \int \frac{A}{P}\,dP + \int \frac{B}{(M-P)}\,dP = \frac{1}{M}\int \frac{1}{P}\,dP - \frac{1}{M}\int \frac{-1}{(M-P)}\,dP$

$= \frac{1}{M} ln(P) - \frac{1}{M} ln(M-P) = \frac{1}{M} ln(\frac{P}{M-P})$

Therefore, $\frac{1}{M} ln(\frac{P}{M-P}) = kt + C \Rightarrow ln(\frac{P}{M-P}) = Mkt + MC$

$\Rightarrow \frac{P}{M-P} = e^{MC}e^{Mkt} \Rightarrow P = Me^{MC}e^{Mkt} - Pe^{MC}e^{Mkt}$

$\Rightarrow P(1 + e^{MC}e^{Mkt}) = Me^{MC}e^{Mkt}$

Then we have $P(t) = \frac{Me^{MC}e^{Mkt}}{(1 + e^{MC}e^{Mkt})}$. $So\ P(0) = \frac{Me^{MC}}{(1 + e^{MC})} = P_0$

$\Rightarrow C = \frac{1}{M} ln(\frac{P_0}{M-P_0})$

Therefore, $P(t) = \frac{(\frac{MP_0}{M-P_0})e^{Mkt}}{(1 + (\frac{P_0}{M-P_0})e^{Mkt})}$ is the model for the population at time $t \geq 0$, after plugging in for C and simplifying.

Note that the $\lim_{t \to \infty} P(t) = \frac{Mk(\frac{MP_0}{M-P_0})e^{Mkt}}{Mk(\frac{P_0}{M-P_0})e^{Mkt}}$ = M (using L'Hospital's rule), meaning that the population approaches its carrying capacity as time goes on.

EXERCISES:

Solve the following first order differential equations.

(1.) $y^2e^{-2x} = \frac{dy}{dx}$, with y(0) = 1.

(2.) $\frac{x^2+3}{e^y} = \frac{dy}{dx}$, with y(0) = 2.

(3.) $(y+1)e^x\,dx + dy = 0$.

(4.) $e^{-x^2}\,dx = \frac{1}{2x}\,dy$, with y(0) = -1

(#) Exact Differential Equations: When we have a function of two variables f(x,y), the differential of f(x,y) is: df = ($\frac{df}{dx}$)dx + ($\frac{df}{dy}$)dy. When we write $\frac{df}{dx}$, where f is a function of two variables x and y, we mean that x should be considered the variable, and y should be treated as if it were a constant. The situation is the same for $\frac{df}{dy}$, y is the variable and x should be treated as if it were a constant. For example,

$$\frac{d}{dx}(x^2y + ysin(x)) = 2yx + ycos(x)$$

and, $\frac{d}{dy}(x^2y + ysin(x)) = x^2 + sin(x)$.

If f(x,y) is continuous and the derivatives $\frac{df}{dx}$ and $\frac{df}{dy}$ are continuous throughout a convex region of the x-y plane, then the so-called mixed derivatives $\frac{d}{dx}(\frac{df}{dy}) = \frac{d^2f}{dx\,dy}$ and $\frac{d}{dy}(\frac{df}{dx}) = \frac{d^2f}{dy\,dx}$ are equal.

So if we have a differential equation M(x,y)dx + N(x,y)dy = 0, and $\frac{dM}{dy} = \frac{dN}{dx}$,
then we should search for a solution of the form f(x,y) = C where the two conditions M(x,y) = $\frac{df}{dx}$ and N(x,y) = $\frac{df}{dy}$ are true. If $\frac{dM}{dy} = \frac{dN}{dx}$, then we have what we call an exact differential equation. Let's illustrate how such an equation can be solved with some examples:

<u>Example 1:</u> Solve $(ysin(x) + xy)\ dx + (\frac{1}{2}x^2 - cos(x) + 2)\ dy = 0$.

First we test to see if we have an exact equation:
Since $\frac{dM}{dy} = (sinx + x)$ and $\frac{dN}{dx} = (sinx + x)$, which says that $\frac{dM}{dy} = \frac{dN}{dx}$,
then we do have an exact differential equation.

Now, since M = $\frac{df}{dx}$, then f(x,y) = $\int M\ dx$.

So f(x,y) = $\int(ysin(x) + xy)\ dx = \ -ycos(x) + (y)\frac{x^2}{2} + h(y)$.

The constant of integration can be a function of y, since y is considered to be constant when integrating with respect to the variable x.

So now, $\frac{df}{dy} = \frac{1}{2}x^2 - cos(x) + h'(y) \ = \ N(x,y) \ = \frac{1}{2}x^2 - cosx + 2$.
This says that h'(y) = 2 $\Rightarrow$ h(y) = 2y + C.

Therefore, f(x,y) = $\frac{1}{2}yx^2 - ycos(x) + 2y + C$ is the solution.

<u>Example 2</u>: Solve $(x^2y^2 - xy + 3x)\ dx + (\frac{2}{3}yx^3 - \frac{1}{2}x^2)\ dy = 0$.

Since $\frac{dM}{dy} = 2x^2y - x = \frac{dN}{dx}$, the equation is exact.

So, $\frac{df}{dx} = M(x,y) = x^2y^2 - xy + 3x$
Therefore, f(x,y) = $\int M(x,y)\ dx = \frac{1}{3}y^2x^3 - \frac{1}{2}yx^2 + \frac{3}{2}x^2 + h(y)$.

Then $\frac{df}{dy} = \frac{2}{3}x^3y - \frac{1}{2}x^2 + h'(y) \;\; = \;\; N(x,y) = \frac{2}{3}yx^3 - \frac{1}{2}x^2$.
So, h'(y) = 0, or h(y) = C.

Therefore, f(x,y) = $\frac{1}{3}y^2x^3 - \frac{1}{2}yx^2 + \frac{3}{2}x^2 + C$ is the solution.

Exact differential equations have many applications in math, science, and engineering, mostly of a more advanced nature than what we are concerned with in this book. However, one application that is within the scope of this book, and important to what we are doing here is finding the solution to Linear First Order Differential Equations, the subject of the next section.

<u>EXERCISES</u>:

Verify that each of the following differential equations is exact and solve them.

(1.) $(x^3y + 2xy) + (\frac{x^4}{4} + x^2 + 1)\frac{dy}{dx} = 0$

(2.) $(e^xy^2 - y^2 + 3)\,dx + (2ye^x - 2xy + 2)\,dy = 0$

(3.) $(x\,siny - xy^2)\,dx + (\frac{1}{2}x^2cosy - yx^2)\,dy = 0$

(#) <u>Linear First Order Differential Equations</u>: A linear first order ODE is one of the form $\frac{dy}{dx} + p(x)y = q(x)$. The way that this equation is usually solved is by multiplying through, on both sides, by a so-called integrating factor $I(x)$ which changes it to an exact ODE which we know how to solve. That is, an ODE in the form $\frac{dy}{dx} + p(x)y = q(x)$ can be rearranged to the form

$$(p(x)y - q(x))\, dx + dy = 0 \text{, which can be made exact.}$$

We'll derive a formula which can be used to find an integrating factor $I(x)$. Multiplying through by $I(x)$ leads to

$$I(x)(p(x)y - q(x))\, dx + I(x)\, dy = 0 .$$

Then if this is actually exact, we must have

$$\frac{d}{dy}(I(x)(p(x)y - q(x))) = \frac{d}{dx}(I(x)) .$$

This says, since $I(x)$, p(x), and q(x) are functions only of x, that $I(x)p(x) = \frac{dI(x)}{dx}$. Note, this is variables separable. So that we have $\frac{dI(x)}{I(x)} = p(x)\, dx \quad \Rightarrow \quad \int \frac{1}{I(x)}\, dI(x) = \int p(x)\, dx$. So $ln|I(x)| = \int p(x)\, dx$.

Then $I(x) = e^{\int p(x)\, dx}$ is our integrating factor.

Example 1: Solve $\frac{dy}{dx} + xy = -x$.

The integrating factor $I(x) = e^{\int p(x)\, dx} = e^{\int x\, dx} = e^{\frac{x^2}{2}}$.

Multiplying both sides of the differential equation by $e^{\frac{x^2}{2}}$ yields

$$(e^{\frac{x^2}{2}})\frac{dy}{dx} + xe^{\frac{x^2}{2}}y = -xe^{\frac{x^2}{2}} .$$

Note that the left-hand side is in the form $\frac{d}{dx}(ye^{\frac{x^2}{2}})$, which is the form $\frac{d}{dx}(y \cdot I(x))$. This will always be the case.

So we have, $\frac{d}{dx}(ye^{\frac{x^2}{2}}) = -xe^{\frac{x^2}{2}}$. This can be rearranged,

$d(ye^{\frac{x^2}{2}}) = -xe^{\frac{x^2}{2}}\,dx \quad \Rightarrow \quad \int d(ye^{\frac{x^2}{2}}) = -\int xe^{\frac{x^2}{2}}\,dx$

$\Rightarrow \quad ye^{\frac{x^2}{2}} = -e^{\frac{x^2}{2}} + C$. Therefore, the solution is $y = Ce^{-\frac{x^2}{2}} - 1$.

Example 2: Solve $\frac{dy}{dx} + (\frac{1}{x})y = \frac{4}{x}$, subject to the condition y(3) = 2.

$I(x) = e^{\int p(x)\,dx} = e^{\int \frac{1}{x}\,dx} = e^{ln(x)} = x$.

So multiplying the differential equation through by $I(x)$ leads to

$x\frac{dy}{dx} + y = 4 \quad \Rightarrow \quad \frac{d}{dx}(y \cdot I(x)) = 4 \quad \Rightarrow \quad d(xy) = 4\,dx$

$\Rightarrow \quad \int d(xy) = \int 4\,dx \quad \Rightarrow \quad xy = 4x + C$.

Therefore, $y = 4 + \frac{C}{x}$ is the general solution.

Since y(3) = 2, we have, $2 = 4 + \frac{C}{3} \quad \Rightarrow \quad C = -6$. So the particular solution is $y = 4 - \frac{6}{x}$.

We can always check to see if we have a correct solution by computing the derivatives of y(x) and plugging them, along with y(x), back into the differential equation. Let's do this for this example.

Now, we had $\frac{dy}{dx} + (\frac{1}{x})y = \frac{4}{x}$, and we found the solution to be $y = 4 - \frac{6}{x}$.
So, $\frac{dy}{dx} = \frac{6}{x^2}$. So, $\frac{6}{x^2} + (\frac{1}{x})(4 - \frac{6}{x}) = \frac{4}{x}$. This says that $\frac{6}{x^2} + \frac{4}{x} - \frac{6}{x^2} = \frac{4}{x}$, which is a true statement. So we know that we have found the correct solution.

EXERCISES:

Solve the following differential equations.

(1.) $\frac{dy}{dx} + 3y = x$

(2.) $\frac{dy}{dx} + y = e^{2x}$

(3.) $\frac{dy}{dx} + (\frac{1}{7x})y = \frac{1}{x^3}$

(13) Second Order Differential Equations

There is so much to the theory and application of first and second order ODE's, but it is not necessary to what we are trying to accomplish here in this part of the book, which is to show the reader how differential and integral calculus can be applied. Therefore we will limit the discussion at this point to just the one following type of second order ODE situation, which actually has much application.

(#) Linear Homogeneous Second Order ODE's with Constant Coefficients:

A differential equation of the form $a\frac{d^2y}{dx^2} + b\frac{dy}{dx} + cy = 0$ is linear in its derivatives, homogeneous (because the right-hand side is 0), and has constant coefficients a, b, and c. These types of differential equations actually have important applications. The way that we solve these types of equations is by writing what is called an auxiliary equation $aD^2 + bD + c = 0$. D^n represents an nth order derivative and c (by itself) means c multiplied by the unknown function y(x), which is what we seek.

We are looking for solutions of the form $y(x) = Ce^{\lambda x}$, where λ is a constant. A second order equation will have two linearly independent solutions involving two unknown constants in its general solution, just as a first order equation has one unknown constant as part of its general solution.

Case (I) If $aD^2 + bD + c = 0$ is the auxiliary equation, and D = λ_1, λ_2 are two distinct real solutions, then the general solution is y(x) = $C_1e^{\lambda_1 x} + C_2e^{\lambda_2 x}$.

Case (II) If $aD^2 + bD + c = 0$ is the auxiliary equation, and λ_1, λ_1 are two repeated real solutions, then the general solution is y(x) = $C_1e^{\lambda_1 x} + C_2xe^{\lambda_1 x}$. These two solutions are linearly independent.

Case (III) If $aD^2 + bD + c = 0$ is the auxiliary equation, and it has two complex conjugate solutions $\lambda_1 \pm \lambda_2 i$, then the general solution is $y(x) = C_1e^{(\lambda_1 + \lambda_2 i)x} + C_2e^{(\lambda_1 - \lambda_2 i)x}$.

There is a formula known as Euler's Identity that is useful here:
$e^{i\lambda x} = cos(\lambda x) + i \cdot sin(\lambda x)$ and
$e^{-i\lambda x} = cos(\lambda x) - i \cdot sin(\lambda x)$ are its two formulations.

The solution above: $y(x) = C_1e^{(\lambda_1 + \lambda_2 i)x} + C_2e^{(\lambda_1 - \lambda_2 i)x}$ equals
$y(x) = C_1e^{\lambda_1 x}(cos(\lambda_2 x) + i \cdot sin(\lambda_2 x))$ + $C_2e^{\lambda_1 x}(cos(\lambda_2 x) - i \cdot sin(\lambda_2 x))$, or
$y(x) = (C_1 + C_2)e^{\lambda_1 x} \cdot cos(\lambda_2 x)$ + $(i \cdot (C_1 - C_2))e^{\lambda_1 x} \cdot sin(\lambda_2 x)$.

We can just let $(C_1 + C_2) = C_1$, and let $(i \cdot (C_1 - C_2)) = C_2$.
So $y(x) = (C_1)e^{\lambda_1 x} \cdot cos(\lambda_2 x) + (C_2)e^{\lambda_1 x} \cdot sin(\lambda_2 x)$ is the general solution.

Often we have initial conditions y(0) = A and y'(0) = B, which allow us to find the unknown constants C_1 and C_2.

Example 1: Solve $2\frac{d^2y}{dx^2} + 4\frac{dy}{dx} + 6y = 0$, with the initial conditions y(0) = 4, y'(0) = -3.

The auxiliary equation is $2D^2 + 4D + 6 = 0$, which has solutions D = $\frac{-4 \pm \sqrt{16-(4)(2)(6)}}{4}$ = $\frac{-4 \pm \sqrt{-32}}{4}$ = $\frac{-4 \pm \sqrt{16}\sqrt{2} \cdot i}{4}$ = $\{-1 + (\sqrt{2})i, \ -1 - (\sqrt{2})i\}$, and this leads to the general solution:
$y(x) = (C_1)e^{-x} \cdot cos(\sqrt{2}x) + (C_2)e^{-x} \cdot sin(\sqrt{2}x)$.

Now, $y'(x) = C_1(e^{-x})(cos(\sqrt{2}\,x) + \sqrt{2}\,sin(\sqrt{2}\,x)) + C_2(e^{-x})(\sqrt{2}\,cos(\sqrt{2}\,x) - sin(\sqrt{2}\,x))$ along with the two initial conditions leads to $C_1 = 4$, and $C_2 = \frac{-7}{\sqrt{2}}$.

So, $y(x) = (4)e^{-x} \cdot cos(\sqrt{2}x) + (\frac{-7}{\sqrt{2}})e^{-x} \cdot sin(\sqrt{2}x)$ is the particular solution that satisfies the initial conditions.

Example 2: Solve $\frac{d^2y}{dx^2} - 4\frac{dy}{dx} + 4y = 0$, subject to y(0) = 1, y'(0) = -1.

The auxiliary equation is $D^2 - 4D + 4 = 0$ has two repeated roots {2, 2}. This leads to the general solution $y(x) = C_1e^{2x} + C_2xe^{2x}$.

Now, y'(x) = $2C_1e^{2x} + C_2(e^{2x})(2x+1)$, and along with the initial conditions, this leads to $C_1 = 1$ and $C_2 = -3$. So the particular solution to this initial value problem is $y(x) = e^{2x} - 3xe^{2x}$.

Example 3: Find the general solution to $\frac{d^2y}{dx^2} + 4\frac{dy}{dx} - y = 0$.

The auxiliary equation is $D^2 + 4D - 1 = 0$. It has two distinct real solutions: $\{-2 + \sqrt{5}, \ -2 - \sqrt{5}\}$. Therefore, the general solution is:

$$y(x) = C_1 e^{(-2 + \sqrt{5})x} + C_2 e^{(-2 - \sqrt{5})x}.$$

In the next two sections, we will discuss two applications of the type of differential equation that we have just considered, that is, a homogeneous second order linear differential equation with constant coefficients. The first application is known as Hooke's Law, which is usually discussed in a first course in Physics when considering a system where we have a mass attached to a spring which exhibits oscillatory motion. The second application is also an elementary physics problem, the analysis of a pendulum exhibiting oscillatory motion. Both of these physical systems are analyzed using the same type of differential equation.

(#) <u>Hooke's Law</u>: This law is usually formulated as F = -kx, where k is a positive constant. We attach a mass m to a spring, with the spring attached to a wall and the mass-spring system sitting in a horizontal way on a table. Then stretch the spring a certain distance A and release the mass (at time t = 0). Assuming a frictionless table and no internal friction for the spring, the resulting motion should be oscillatory or simple harmonic motion. The above mathematical equation says that the restoring force F(t) of the spring is directly proportional to the displacement x(t) and acting in the opposite direction of the displacement because of the negative sign. The displacement x(t) is measured from the resting equilibrium position. "k" is called the spring constant. Since we know Newton's Law that F = ma, this equation can be written:

ma = -kx, or as a differential equation, $m\frac{d^2x}{dt^2} = -kx(t)$

We have the initial conditions x(0) = A, and x'(0) = 0, since the spring is stretched an initial amount A from equilibrium position and the velocity of the mass is 0 at the instant it is released. So we have the auxiliary equation

$$mD^2 + k = 0 \quad \Rightarrow \quad D^2 = \frac{-k}{m} \quad \Rightarrow \quad D = \pm(\sqrt{\tfrac{k}{m}})i$$

So we have x(t) = $C_1 e^{i(\sqrt{\frac{k}{m}})t} + C_2 e^{-i(\sqrt{\frac{k}{m}})t}$, or

$$\begin{aligned} x(t) &= C_1(cos(\sqrt{\tfrac{k}{m}}t) + i\,sin(\sqrt{\tfrac{k}{m}}t)) + C_2(cos(\sqrt{\tfrac{k}{m}}t) - i\,sin(\sqrt{\tfrac{k}{m}}t)) \\ &= (C_1 + C_2)cos(\sqrt{\tfrac{k}{m}}t) + i\,(C_1 - C_2)sin(\sqrt{\tfrac{k}{m}}t) \\ &= C_1\,cos(\sqrt{\tfrac{k}{m}}t) + C_2\,sin(\sqrt{\tfrac{k}{m}}t) \end{aligned}$$

From x(0) = A $\Rightarrow$ A = $C_1 cos(0) + C_2 sin(0)$ $\Rightarrow$ $C_1 = A$.

$$x'(t) = -C_1(\sqrt{\tfrac{k}{m}})sin(\sqrt{\tfrac{k}{m}}t) + C_2(\sqrt{\tfrac{k}{m}})cos(\sqrt{\tfrac{k}{m}}t).$$

So from x'(0) = 0 $\Rightarrow$ $0 = -C_1(\sqrt{\tfrac{k}{m}})\,sin(0) + C_2(\sqrt{\tfrac{k}{m}})\,cos(0)$

$\Rightarrow$ $C_2 = 0$.

So we have that the position at time t is x(t) = $A\,cos(\sqrt{\tfrac{k}{m}}t)$.
Assuming no frictional forces are involved in the motion, the mass will oscillate back and forth according to this position function x(t).

$$x'(t) = -A(\sqrt{\tfrac{k}{m}})\,sin(\sqrt{\tfrac{k}{m}}t)$$
$$x''(t) = -A(\tfrac{k}{m})\,cos(\sqrt{\tfrac{k}{m}}t)$$

Setting x'(t) = 0 $\Rightarrow$ t = 0 is the critical value where x(t) will be minimized or maximized. Since x''(0) = $-A(\frac{k}{m}) < 0$, x(t) reaches a maximum distance of A units away from the equilibrium position as the mass oscillates back and forth. A is called the amplitude of the motion.

Setting x″(t) = 0 $\Rightarrow$ $cos(\sqrt{\frac{k}{m}}t) = 0$ $\Rightarrow$ $\sqrt{\frac{k}{m}}t = \frac{\pi}{2}$

$\Rightarrow$ t = $\frac{\pi}{2}(\sqrt{\frac{m}{k}})$ is the critical value where x′(t) is minimized or maximized.

$|x'(\frac{\pi}{2}(\sqrt{\frac{m}{k}}))| = |-A(\sqrt{\frac{k}{m}})\, sin(\frac{\pi}{2})| = (A\sqrt{\frac{k}{m}})$, and x′(0) = 0.

x($\frac{\pi}{2}\sqrt{\frac{m}{k}}) = 0$, and x(0) = A.

From this we can see that the velocity is a maximum of (A $\sqrt{\frac{k}{m}}$) at time t = $(\frac{\pi}{2}\sqrt{\frac{m}{k}})$, which is as the mass passes through the equilibrium position. The velocity is at a minimum of 0 when the mass is A units away from the equilibrium position because it stops for an instant at that point of its motion.

(#) <u>Pendulum Motion</u>: When we have a mass m suspended from a nearly internally frictionless wire of length L, and if it is oscillating back and forth influenced only by the Earth's gravity (after having been set into motion by releasing it from the non-equilibrium position), then we have for small angles θ a system which exhibits to a very high degree simple harmonic motion. So this system is very much like the mass-spring system of the previous section and we can analyze its motion in an analogous way. For the mass m moving in a circular arc of radius L, the arc length s = Lθ. So $\frac{d^2s}{dt^2} = L\frac{d^2\theta}{dt^2}$. This is essentially the acceleration of the system in the horizontal direction x. Notice that we are saying the arc length s $\approx$ the horizontal projection of the mass at x, for small θ. Looking at things differently, this acceleration is caused by the horizontal component of the Earth's gravity acting on the mass m and is given by (-g sin θ). See the diagram below. Equating the two accelerations, we have for small angles θ the differential equation $L\theta''(t) = -g sin\theta$. Because θ is small (usually thought of as being about 15° or less),

$\sin\theta \approx \theta$, so we have the differential equation $L\theta''(t) = -g\theta(t)$. This second order linear and homogeneous ODE with constant coefficients has the exact same form as the ODE used in Hooke's Law.

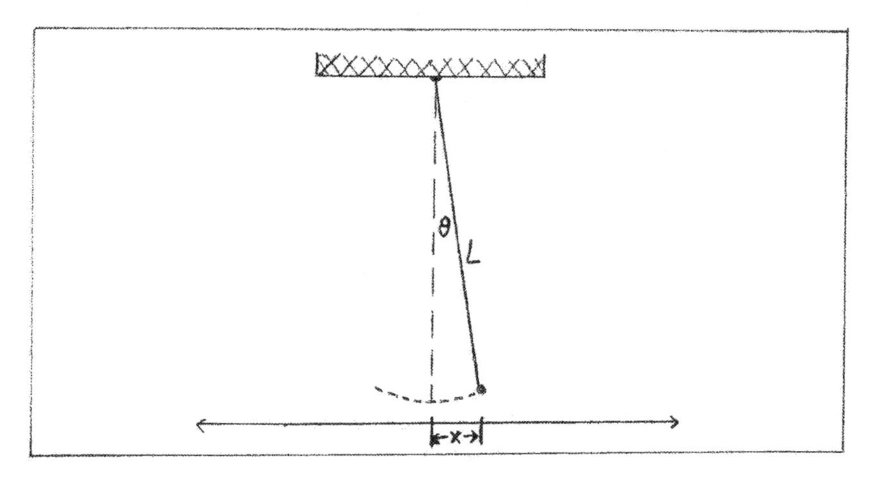

The auxiliary equation is $LD^2 + g = 0$. So $D^2 = \frac{-g}{L}$, and D = $\pm\sqrt{\frac{g}{L}}\,(i)$.
When we consider the two initial conditions $\theta(0) = A$, and $\theta'(0) = 0$, this leads to the solution $\theta(t) = A\,cos(\sqrt{\frac{g}{L}}\,t)$, completely analogous to the differential equation analysis of the previous section on Hooke's Law. Note that A is a maximum angular displacement, not a distance in the x-direction.

As a note, if we do want an analysis of pendulum motion in the x-direction then note that since x = L$\sin\theta \approx$ Lθ for small θ, so that $\frac{d^2x}{dt^2} \approx L\frac{d^2\theta}{dt^2}$. Then the above equation $L\theta''(t) = -g\theta(t)$ can be written $L(L\theta''(t)) = -g(L\theta(t))$ which says that $L\frac{d^2X}{dt^2} = -gx(t)$. This equation is in the same form as the one involving θ. Therefore, we could do the entire analysis in terms of the x-component of the motion and the results would be entirely the same.

Getting back to the variable θ, this simple harmonic motion can be thought of as $\theta(t) = A\,cos(\omega t)$, where ω is called the angular velocity. Since $\omega = 2\pi f$ (where f is the frequency of the motion) and since $f = \frac{1}{T}$ (where T is the period of the motion), we have $\omega = \frac{2\pi}{T}$. So the period of the pendulum's swing, which is the time that it takes to complete one cycle, is

$T = \frac{2\pi}{\omega}$. $\omega = \sqrt{\frac{g}{L}}$, so $T = 2\pi\sqrt{\frac{L}{g}}$. Note that the period is independent of mass m.

Velocity for the angle θ is given by $\theta'(t) = -A(\sqrt{\frac{g}{L}})\, sin(\sqrt{\frac{g}{L}}\, t)$. This is 0 when t = 0 (one of our initial conditions), which is at the instant the mass is released from a maximum angular displacement A. Velocity is a maximum when t = $(\frac{\pi}{2}\sqrt{\frac{L}{g}})$, which is when the mass is in the equilibrium position, and this maximum velocity for θ is $(A\sqrt{\frac{g}{L}})$. So the maximum angular velocity decreases and the period increases as L increases.

EXERCISES:

Solve the following differential equations.

(1.) $2\frac{d^2y}{dx^2} + 4\frac{dy}{dx} - 8y = 0$

(2.) $\frac{d^2y}{dx^2} + 20y = 0$

(3.) $\frac{d^2y}{dx^2} - 2\frac{dy}{dx} + y = 0$, where y(0) = 0, y'(0) = 1

(4.) $\frac{d^2y}{dx^2} - \frac{dy}{dx} - 6y = 0$, where y(0) = 2, y'(0) = 5

Appendix: Answers to Exercises

PART I: NUMBERS, SEQUENCES & SERIES, LIMITS & CONTINUITY

CHAPTER (3) LIMITS AND CONTINUITY

Exercises: Limits and Continuity

(1.) 9 (2.) 0 (3.) 1 (4.) $+\infty$ (5.) $-\infty$ (6.) $\frac{1}{8}$

(7.) $\frac{3}{5}$ (8.) $-\frac{2}{5}$ (9.) $-\infty$ (10.) $-\infty$ (11.) $+\infty$

(12.) $-\infty$ (13.) $\frac{1}{2}$ (14.) 0 (15.) 1 (16.) $\frac{1}{4}$ (17.) 0

(18.) $-\infty$ (19.) x = 1 (20.) x = -1,1 (21.) x = -2,4

(22.) continuous everywhere since the denominator cannot be 0.

PART II: DIFFERENTIAL CALCULUS

CHAPTER (4) THE DERIVATIVE

Exercies: Calculating Derivatives from the Definition

(1.) y'(x) = 10x + 2 (2.) y'(x) = $\frac{-3}{x^2}$. (3.) y'(x) = $\frac{1}{2\sqrt{x}}$.

CHAPTER (5) DERIVATIVES OF ALGEBRAIC FUNCTIONS

Exercises: Derivatives of Algebraic Functions

(1.) $\frac{dy}{dx} = 9x^2 - 12x$ (2.) $\frac{dy}{dx} = \frac{x}{\sqrt{x^2-4}}$ (3.) $\frac{dy}{dx} = \frac{3\sqrt{x}}{2} - \frac{9}{2\sqrt{x}}$

(4.) $\frac{dy}{dx} = \frac{1-2xy^2}{2x^2y}$ (5.) $\frac{dy}{dx} = \frac{-45}{x^4}$ (6.) $\frac{dy}{dx} = \frac{2x-3}{2y}$

(7.) $\frac{dy}{dx} = (x-8)^3\left(\frac{9x-8}{2\sqrt{x}}\right)$ (8.) $\frac{dy}{dx} = \frac{1}{8}x^{\frac{-7}{8}}$ (9.) $\frac{dy}{dx} = \frac{2x^{-\frac{1}{3}}}{3} - \frac{x^{-\frac{4}{3}}}{3}$

(10.) $\frac{dy}{dx} = \frac{-2}{2y-1}$ (11.) $\frac{dy}{dx} = \frac{-x}{y}$ (12.) $\frac{dy}{dx} = 2x+1$

(13.) $\frac{dy}{dx} = 8x(x^2-16)^3$ (14.) $\frac{dy}{dx} = \frac{1}{5}(x^6+x^4)^{-\frac{4}{5}}(6x^5+4x^3)$

(15.) $\frac{dy}{dx} = 90x^4(x^5+4)^8$ (16.) $\frac{dy}{dx} = \frac{1}{2}$

(17.) For $y = x^2+3$, $\frac{dy}{dx}$ = 2x, $\frac{dx}{dy} = \frac{1}{2x}$ (reciprocals).

(18.) y = $\frac{1}{\sqrt{5}}x + (\sqrt{5})$; The slope $\frac{1}{\sqrt{5}}$ is the instantaneous rate of change of y = $\sqrt{4x}$ with respect to x when x = 5.

(19.) y = -3x + $\frac{21}{4}$; The slope -3 is the instantaneous rate of change of y = $3-x^2$ with respect to x when x = $\frac{3}{2}$.

(20.) y'(1) = -3, y'(5) = $\frac{-3}{25}$, y'(10) = $\frac{-3}{100}$; The rate of change of y = $\frac{3}{x}$ with respect to x is negative and converging to 0 as $x \to \infty$.

(21.) y'($\frac{1}{2}$) = -12, y'($\frac{1}{5}$) = -75, y'($\frac{1}{10}$) = -300 ; The rate of change of y = $\frac{3}{x}$ with respect to x is negative and approaching ($-\infty$) as x approaches 0 from the right.

CHAPTER (6) DERIVATIVES OF TRANSCENDENTAL FUNCTIONS

Exercises: Derivatives of Transcendental Functions

(1.) $\frac{dy}{dx} = (3)e^{(3x+1)}$ (2.) $\frac{dy}{dx} = \frac{2}{x}$ (3.) $\frac{dy}{dx} = \frac{8x}{(ln4)(4x^2+2)}$

(4.) $\frac{dy}{dx} = 5^x(ln5)$ (5.) $\frac{dy}{dx} = e^{-x}(2x-x^2)$ (6.) $\frac{dy}{dx} = \frac{(1-3ln(x))}{x^4}$

(7.) $\frac{dy}{dx} = cos(\sqrt{x}) \cdot \frac{1}{2\sqrt{x}}$ (8.) $\frac{dy}{dx} = (-2) \cdot (cos(2x)) \cdot (sin(sin(2x)))$

(9.) $\frac{dy}{dx} = sec^2(x - cos(x)) \cdot (1 + sin(x))$ (10.) $\frac{dy}{dx} = (\csc(x)) \cdot (1 - x(cot(x)))$

(11.) $\frac{dy}{dx} = (\ln(2)) \cdot (6x+4) \cdot 2^{(3x^2+4x+2)}$ (12.) 0

(13.) $\frac{dy}{dx} = (4^{-x})(1-(ln4)x)$ (14.) $\frac{dy}{dx} = \frac{-x}{\sqrt{2\pi}} \cdot e^{\frac{-x^2}{2}}$ (15.) $\frac{dy}{dx}$ = sin(2x)

(16.) $\frac{dy}{dx}$ = (9) $(tan(3x)) \cdot (sec^3(3x))$ (17.) $\frac{dy}{dx} = \frac{1}{\sqrt{1-(x+4)^2}}$

(18.) $\frac{dy}{dx} = \frac{7}{(x)\sqrt{100x^{14}-1}}$ (19.) $\frac{dy}{dx} = \frac{2e^{2x}}{1+e^{4x}}$ (20.) $\frac{dy}{dx} = \frac{(-4x)}{1+4x^4}$

(21.) $y = (3e^6)x - (5e^6)$. The slope ($3e^6$) is the instantaneous rate of change of y = e^{3x} with respect to x when x = 2.

(22.) $y = \frac{1}{6}x + (ln(6) - \frac{7}{6})$. The slope ($\frac{1}{6}$) is the instantaneous rate of change of $y = ln(x-1)$ with respect to x when x = 7.

(23.) y′(0) = 1, y′($\frac{\pi}{6}$) = $\frac{\sqrt{3}}{2}$, y′($\frac{\pi}{3}$) = $\frac{1}{2}$; This indicates that the instantaneous rate of change of y = sin(x) with respect to x is decreasing to 0 as x $\rightarrow \frac{\pi}{2}$ from the left.

CHAPTER (7) APPLICATIONS OF THE DERIVATIVE

Exercises: Curve Analysis

(1.) $(y = \frac{1}{x^2})$

(a) vertical asymptote at x = 0
(b) horizontal asymptote at y = 0
(c) y′(x) = $\frac{-2}{x^3}$, critical value: 0 , y″(x) = $\frac{6}{x^4}$, inflection value: 0
(d) increasing: $(-\infty, 0)$
(e) decreasing: $(0, \infty)$
(f) concave up: $(-\infty, 0) \cup (0, \infty)$
(g) concave down: nowhere
(h) no local minimums or maximums
(i) no absolute minimums or maximums
(j) no x or y intercepts

(2.) $(y = sin(x),\ 0 < x < 2\pi)$

(a) no vertical asymptotes
(b) no horizontal asymptotes

(c) y'(x) = cos(x), critical values: $\{\frac{\pi}{2}, \frac{3\pi}{2}\}$,

y''(x) = -sin(x), inflection values: $\{\pi\}$

(d) increasing: $(0, \frac{\pi}{2}) \cup (\frac{3\pi}{2}, 2\pi)$

(e) decreasing: $(\frac{\pi}{2}, \frac{3\pi}{2})$

(f) concave up: $(\pi, 2\pi)$

(g) concave down: $(0, \pi)$

(h) local max: 1 at x = $\frac{\pi}{2}$, local min: -1 at x = $\frac{3\pi}{2}$

(i) absolute min: -1 at x = $\frac{3\pi}{2}$, absolute max: 1 at x = $\frac{\pi}{2}$

(j) x-intercept: π , no y-intercept

(3.) $(y = tan(x), \ -\frac{\pi}{2} < x < \frac{\pi}{2})$

(a) vertical asymptotes: x = $\frac{-\pi}{2}$, $x = \frac{\pi}{2}$.

(b) horizontal asymptote: none

(c) y'(x) = $sec^2(x)$, critical values: none

y''(x) = 2($sec^2(x)tan(x)$, inflection value: 0

(d) increasing on $(\frac{-\pi}{2}, \frac{\pi}{2})$

(e) decreasing nowhere.

(f) concave up: $(0, \frac{\pi}{2})$

(g) concave down: $(\frac{-\pi}{2}, 0)$

(h) no local minimums or maximums

(i) no absolute minimums or maximums

(j) x-intercept: 0 , y-intercept: 0

(4.) $(y = e^{-x})$

(a) vertical asymptotes: none

(b) horizontal asymptote: y = 0 as $x \rightarrow \infty$

(c) y'(x) = $-e^{-x}$, no critical values

y''(x) = e^{-x}, no inflection values

(d) never increasing

(e) always decreasing

(f) concave up everywhere

(g) concave down nowhere

(h) no local minimums or maximums

(i) no absolute minimums or maximums

(j) x-intercepts: none , y-intercepts: 1

Exercises: Optimization and Related Rates

(1.) The maximum value of S is $(\frac{17}{2})$, when x = $\frac{1}{2}$ and y = $\frac{1}{2}$.

(2.) The maximum value of P is 4624, when x = 68, y = 68.

(3.) The minimum value of surface area is approximately 625.13,

when r = $\sqrt[3]{\frac{600}{\pi}} \approx 5.76$, and h = $\frac{1200}{\pi(\sqrt[3]{\frac{600}{\pi}})^2} \approx 11.52$.

(4.) $\frac{dV}{dt} \approx$ 458.67 $ft.^3$/minute.

PART III: INTEGRAL CALCULUS

CHAPTER (8) INDEFINITE INTEGRALS

Exercises: Integration with a Change of Variable

(1.) $\frac{-1}{6}\, cos(3x^2) + C$ (2.) $\frac{1}{72}\,(4x^3 - 9)^6 + C$

(3.) $\frac{1}{5}\, ln|sec(5x) + tan(5x)| + C$ (4.) $\frac{1}{7}\, ln|x| + C$ (5.) $\frac{1}{13}\, e^{13x} + C$

(6.) $\frac{1}{26}\, e^{(13x^2+1)} + C$ (7.) $\frac{1}{14}\, ln|7x^2 - 12| + C$ (8.) $\frac{1}{64}\,(8x^2 - 16x)^4 + C$

(9.) $\frac{1}{3}\,(\sqrt{x} + 2)^6 + C$ (10.) $\frac{1}{60}\,(10x^2 - 20x)^3 + C$

(11.) $\frac{-1}{5}\, ln|cos(5x)| + C$ (12.) $\frac{1}{2}\, sin(x^2 + 2) + C$ (13.) $\frac{1}{4}\, sin^{-1}(4x) + C$

(14.) $\frac{1}{5}\, tan^{-1}(5x) + C$ (15.) $(1 + x^2)^{\frac{3}{2}} + C$ (16.) $\frac{37}{9}\,(1 + 3x^2)^{\frac{3}{2}} + C$

(17.) $\frac{1}{2}\, e^{x^2} + C$ (18.) $\frac{1}{8}\, ln|8x + 2| + C$ (19.) $\frac{-1}{6}\,(3x + 2)^{-2} + C$

(20.) $\frac{1}{5}\,(2x^2 + 1)^{\frac{5}{4}} + C$

Exercises: Integration By Parts and Partial Fractions

(1.) $e^x(x-1)+C$ (2.) $3x\sin(x)+3\cos(x)+C$ (3.) $\frac{2}{3}x^{\frac{3}{2}}(\ln(x)-\frac{2}{3})+C$

(4.) $e^x(x^2-2x+2)+C$ (5.) $-x\cos(4x)+\frac{1}{4}\sin(4x)+C$

(6.) $(x)(\ln(x)-1)+C$ (7.) $\frac{-1}{9}\ln|x|+\frac{10}{9}\ln|x-9|+C$

(8.) $\ln|x|-\ln|x+1|+C$ (9.) $15\ln|x-1|-\frac{15}{2}\ln|x^2+1|+15\tan^{-1}(x)+C$

(10.) $15\tan^{-1}(x)+C$ (11.) $\frac{3x^2}{2}-6x+11\ln|x+1|+\frac{4}{x+1}+C$

(12.) $4x-12\ln|x|+7\ln|x+1|+C$

CHAPTER (9) DEFINITE INTEGRALS

Exercises: Definite Integrals

(1.) $\frac{1}{3}$ (2.) $\frac{16-4\sqrt{2}}{3}$ (3.) 1 (4.) -1 (5.) 8 (6.) $\frac{\pi}{2}$

(7.) $\ln(\sqrt{2}+1)-\ln(\sqrt{3})$ (8.) ≈ 54.7 (9.) 1

(10.) -sin(-2) (11.) ≈ 534.75 (12.) $-\frac{5}{3}$

CHAPTER (10) APPLICATIONS OF INDEFINITE AND DEFINITE INTEGRALS

Exercises: Areas Between Curves

(1.) 20 (2.) $\ln(6)-\frac{5}{6}$ (3.) $\frac{1}{2}(1-e^{-20})$ (4.) ln(2) (5.) $6+2\cos(5)$

(6.) 50 (7.) $\frac{25}{2}$

Exercises: Applications to Physics

(1.) (a) a(t) = $-9.8\ \frac{meters}{second^2}$

v(t) = $300-9.8t\ \frac{meters}{second}$

x(t) = $(-4.9)t^2+300t+200$ meters

(b) 4791.84 meters

(c) 61.88 seconds

(d) 306.42 $\frac{meters}{second}$

(2.) v(t) = $1.5t^2+6.5t$, $t \geq 0$.

x(t) = $0.5t^3 + 3.25t^2$, t ≥ 0.

(3.) (a) 5001.75 Joules

(b) 31.63 $\frac{meters}{second}$.

(4.) 189.752 cm^3

Exercises: Arc Length

(1.) ≈ 1.3170 (2.) ≈ 9.747 (3.) ≈ 0.2869

Exercises: Surfaces of Revolution

(1.) 300($\sqrt{10}$)(π)

(2.) ≈ 7.2116

(3.) 4πr^2 (The reader may recognize that this is the surface area of a sphere)

Exercises: Solids of Revolution

(1.) $\frac{4}{3}\pi r^3$ (The reader may recognize this as the volume of a spherical solid)

(2.) $\frac{\pi}{5}$

(3.) $\frac{6\pi}{5}$

CHAPTER (11) IMPROPER INTEGRALS

Exercises: Improper Integrals

(1.) $-\infty$ (2.) $\frac{3}{\sqrt[3]{2}}$ (3.) $\frac{1}{4}$ (4.) $\frac{1}{50}$ (5.) $+\infty$

PART IV: DIFFERENTIAL EQUATIONS

CHAPTER (12) FIRST ORDER DIFFERENTIAL EQUATIONS

Exercises: Variables Separable Equations

(1.) $y = \frac{1}{\frac{1}{2}e^{-2x} + \frac{1}{2}}$ (2.) $y = ln(\frac{x^3}{3} + 3x + e^2)$

(3.) $e^x + ln|y + 1| = C$ (4.) $y = -e^{-x^2}$

Exercises: Exact Differential Equations

(1.) f(x,y) = $\frac{1}{4}yx^4 + yx^2 + y + C$

(2.) f(x,y) = $y^2e^x - xy^2 + 3x + 2y + C$

(3.) f(x,y) = $\frac{1}{2}siny\, x^2 - \frac{1}{2}x^2y^2 + C$

Exercises: First Order Linear Differential Equations

(1.) $y = \frac{1}{3}x - \frac{1}{9} + Ce^{-3x}$

(2.) $y = \frac{1}{3}e^{2x} + Ce^{-x}$

(3.) $y = \frac{-7}{13x^2} + C$

CHAPTER (13) SECOND ORDER DIFFERENTIAL EQUATIONS

Exercises: Second Order Differential Equations

(1.) y = $C_1e^{(-1+\sqrt{5})x} + C_2e^{(-1-\sqrt{5})x}$

(2.) $y = C_1\, cos(2\sqrt{5}x) + C_2\, sin(2\sqrt{5}x)$

(3.) $y = xe^x + e^x$

(4.) $y = \frac{1}{5}e^{-2x} + \frac{9}{5}e^{3x}$

About The Author:

Timothy C. Kearns graduated from Virginia Polytechnic Institute and State University (with honors) in June of 1983, with a BS degree in Statistics and Mathematics. In addition, he has successfully completed additional graduate level coursework in the field of mathematics, and has been an avid reader of mathematics and science. Since 2003 he has also been a successful private tutor of mathematics, statistics, and physics in the Northern Virginia / Washington D.C. area. He is very passionate about the mathematical sciences, especially calculus and the larger subject of real analysis. He enjoys conveying his expertise in these subjects to high school and college students that are interested in a career in mathematics, engineering, and the mathematically related sciences. This book is a product of his many years of experience with calculus and related subjects, and written for those students that are eager to learn the fundamentals of the differential and integral calculus and its many applications.